W9-CRL-658

A C S   S Y M P O S I U M   S E R I E S **605**

Genetically Modified Foods

Safety Issues

Karl-Heinz Engel, EDITOR
*Bundesinstitut für gesundheitlichen
Verbraucherschutz und Veterinärmedizin*

Gary R. Takeoka, EDITOR
*Agricultural Research Service
U.S. Department of Agriculture*

Roy Teranishi, EDITOR
*Agricultural Research Service
U.S. Department of Agriculture*

Developed from a symposium sponsored
by the Division of Agricultural and Food Chemistry
at the 208th National Meeting
of the American Chemical Society,
Washington, DC,
August 21–25, 1994

American Chemical Society, Washington, DC 1995

Library of Congress Cataloging-in-Publication Data

Genetically modified foods: safety issues / Karl-Heinz Engel, Gary R. Takeoka, Roy Teranishi, editors.

p.   cm.—(ACS symposium series; 605)

"Developed from a symposium sponsored by the Division of Agricultural and Food Chemistry at the 208th National Meeting of the American Chemical Society, Washington, DC, August 21–24, 1994."

Includes bibliographical references and indexes.

ISBN 0–8412–3320–9

1. Food—Biotechnology—Congresses. 2. Crops—Genetic engineering—Congresses.

I. Engel, Karl-Heinz, 1954–   . II. Takeoka, Gary R. III. Teranishi, Roy, 1922–   . IV. American Chemical Society. Division of Agricultural and Food Chemistry. V. American Chemical Society. Meeting (208th: Washington, DC) VI. Series.

TP248.65.F66G46   1995
363.19'2—dc20   95–31995
                                                                            CIP

This book is printed on acid-free, recycled paper.

Foreword

THE ACS SYMPOSIUM SERIES was first published in 1974 to provide a mechanism for publishing symposia quickly in book form. The purpose of this series is to publish comprehensive books developed from symposia, which are usually "snapshots in time" of the current research being done on a topic, plus some review material on the topic. For this reason, it is necessary that the papers be published as quickly as possible.

Before a symposium-based book is put under contract, the proposed table of contents is reviewed for appropriateness to the topic and for comprehensiveness of the collection. Some papers are excluded at this point, and others are added to round out the scope of the volume. In addition, a draft of each paper is peer-reviewed prior to final acceptance or rejection. This anonymous review process is supervised by the organizer(s) of the symposium, who become the editor(s) of the book. The authors then revise their papers according to the recommendations of both the reviewers and the editors, prepare camera-ready copy, and submit the final papers to the editors, who check that all necessary revisions have been made.

As a rule, only original research papers and original review papers are included in the volumes. Verbatim reproductions of previously published papers are not accepted.

Contents

Preface ix

1. **Foods and Food Ingredients Produced via Recombinant DNA**
   Techniques: An Overview 1
   Karl-Heinz Engel, Gary R. Takeoka, and Roy Teranishi

   REGULATORY OVERSIGHT

2. **U.S. Food and Drug Administration Policy for Foods Developed**
   by Biotechnology 12
   J. H. Maryanski

3. **Regulatory Oversight and Safety Assessment of Genetically**
   Modified Foods in the European Union 23
   K.-H. Engel, M. Schauzu, G. Klein, and A. Somogyi

4. **Administrative Oversight To Ensure Safety of Biotechnologically**
   Produced Foods in Japan 33
   Ryoji Takahara

5. **The Role of the Public and Federal Advisory Committees**
   in Providing Advice to the Government on Agricultural Science
   Policy 52
   Alvin L. Young and Daniel D. Jones

6. **Safety Assessment of Flavor Ingredients Produced**
   by Genetically Modified Organisms 59
   J. B. Hallagan and R. L. Hall

   PLANTS

7. **Determination of the Safety of Genetically Engineered Crops** 72
   Keith Redenbaugh, William Hiatt, Belinda Martineau,
   and Donald Emlay

8. Exotic Germ Plasm or Engineered Genes: Comparison of Genetic Strategies To Improve Fruit Quality 88
Alan B. Bennett, Roger Chetelat, and Ellen Klann

9. The Potential for Allergenicity in Transgenic Foods 100
Oscar L. Frick

10. Use of Plant Virus Genes To Produce Disease-Resistant Crops 113
Lynn E. Murry

11. Molecular Cloning of Cereal Cystatins and Evaluation of Their Antiviral and Antipest Effects 124
S. Arai, M. Kuroda, I. Matsumoto, H. Watanabe, and K. Abe

12. Safety Assessment of the *Bacillus thuringiensis* Insecticidal Crystal Protein CRYIA(b) Expressed in Transgenic Tomatoes 134
H. P. J. M. Noteborn, M. E. Bienenmann-Ploum, J. H. J. van den Berg, G. M. Alink, L. Zolla, A. Reynaerts, M. Pensa, and H. A. Kuiper

13. Safety Assessment of Potatoes Resistant to Colorado Potato Beetle 148
P. B. Lavrik, D. E. Bartnicki, J. Feldman, B. G. Hammond, P. J. Keck, S. L. Love, M. W. Naylor, G. J. Rogan, S. R. Sims, and R. L. Fuchs

MICROORGANISMS

14. Improvement of Beer Brewing by Using Genetically Modified Yeast 160
J. Vogel, K. Wackerbauer, and U. Stahl

15. Genetic Modification of Brewer's Yeast To Produce Acetolactate Decarboxylase and the Safety Aspects of the Beer Brewed by the Transformed Yeast 171
R. Takahashi, M. Kawasaki, H. Sone, and S. Yamano

16. Safety Aspects of Genetically Modified Lactic Acid Bacteria 181
W. P. Hammes, C. Hertel, and C. Cavadini

ENZYMES

17. Enzymes from Genetically Modified Microorganisms................... 196
    Sven Pedersen, Bent F. Jensen, and Steen T. Jørgensen

ANIMALS

18. Progress in Genetic Modifications of Farm Animals...................... 210
    V. G. Pursel

INDEXES

Author Index.. 232

Affiliation Index... 232

Subject Index... 233

Preface

BIOTECHNOLOGY IS DEFINED AS THE APPLICATION of scientific and engineering principles to the processing of materials by biological agents to provide goods and services. In 1919, Karl Ereky, an Hungarian agricultural economist, coined the word "biotechnology" to cover the topic of the interaction of biology with technology. In 1979, E. F. Hutton obtained a trademark on the word "biotechnology" to describe a magazine dealing with genetic engineering, and thus the word became associated more with genetic engineering rather than with the more general meaning. Usage and definition of this word are not static and are continuing to evolve. In the symposium upon which this book is based we did not use the word "biotechnology" because of the more general sense as defined originally by Ereky and more recently by the Organization for Economic Cooperation and Development. However, we did use the phrase "genetic engineering" to refer to altering the genetic makeup of organisms.

Just as hybridization helped to recombine plant genes to improve quality and yields of crops, the advent of genetic engineering presents many possibilities for improving acceptability, nutritional values, and yields of conventional crops. The excitement of this new field and the need for such improvements have stimulated tremendous activities in application of genetic engineering to improve farm crops.

The purpose of the symposium was to discuss on a scientific basis some of the various genetic engineering projects involved, directly and indirectly, in the production of foods and flavors and the actual and proposed benefits evolving from such work. Because some critics state that all genetic engineering on food materials must stop, and because it is equally foolish to accept everything that is new as beneficial, the safety aspects of foods resulting from genetically engineered organisms were addressed. Another topic of discussion was governmental guidelines that were installed to protect the consumers by ensuring that the production of greater amounts of food will be done without impairing its nutritional and safety aspects. It seems clear that increased production of food is necessary for feeding the ever-increasing population of this world and that genetic engineering has an important role to play in the needed increase in food production.

This book, composed mostly of papers given at the symposium by international experts in governmental, industrial, and academic

ix

organizations, provides information that should allay the fears of those who question what genetic engineering will do to our foods and to our health. This book should be of interest to scientists involved in those areas of food production in which genetic engineering is active. The anticipated developments in the near future will appeal to consumers who are interested in new food products and in current efforts to keep food safe.

KARL-HEINZ ENGEL
Bundesinstitut für gesundheitlichen
  Verbraucherschutz und Veterinärmedizin
Postfach 330013
D-14191 Berlin
Germany

GARY R. TAKEOKA
ROY TERANISHI
Western Regional Research Center
Agricultural Research Service
U.S. Department of Agriculture
800 Buchanan Street
Albany, CA 94710

June 7, 1995

We dedicate this volume to Martha.

—KHE, GRT, RT—

Chapter 1

Foods and Food Ingredients Produced via Recombinant DNA Techniques

An Overview

Karl-Heinz Engel[1], Gary R. Takeoka[2], and Roy Teranishi[2]

[1]Federal Institute for Health Protection of Consumers and Veterinary Medicine, Postfach 330013, D–14191 Berlin, Germany
[2]Western Regional Research Center, Agricultural Research Service, U.S. Department of Agriculture, 800 Buchanan Street, Albany, CA 94710

Commercial products resulting from recombinant DNA techniques in agricultural biotechnology have just entered or are about to enter the market. Some of the progress in improving nutrition and acceptability of foods as well as improving herbicide, insect, and virus resistance of plants are discussed. Enzymes and micro-organisms can now be more precisely altered to optimize their roles in food production. Genetic modification of animals is much more complex than that of microorganisms and plants, nevertheless, some interesting progress is being made. The applications of genetic engineering in food production are evoking positive and negative reactions. The concepts presently being developed to assess the safety of foods derived from modern biotechnology are outlined. The aspects involved in creating a framework for regulatory oversight on genetically modified foods are also discussed.

The use of microorganisms in the production of foods and beverages, such as bread, cheese, wine, or beer has a long tradition dating back to ancient days. Industrial biotechnology has its roots in the classical production of fermented foods. Therefore, it is a logical consequence that the application of the so-called "modern biotechnology" does not remain limited to areas such as medicine and pharmaceuticals but is increasingly applied in the production of foods and food ingredients. Developments in genetic engineering have created new dimensions in classical biotechnology. By using recombinant DNA techniques, it has become possible to direct the movements of specific and useful segments of genetic material between unrelated organisms, thereby crossing the barriers between plants, animals, and microorganisms. The scope of "modern biotechnology" goes beyond the traditional area of food fermentation processes. It is now possible to make specific genetic modifications in plants and animals that introduce traits or substances that could not be introduced by traditional methods.

0097–6156/95/0605–0001$12.00/0
© 1995 American Chemical Society

After fascinating developments in recent years on the level of basic research, the applications of recombinant DNA techniques in agricultural biotechnology are now heading towards attractive businesses. Commercial products have just entered or are about to enter the market. The spectrum of applications is comprised of genetically modified crops, foods produced from genetically modified microorganisms, as well as food ingredients/additives obtained from genetically modified organisms.

Plants

The enormous advances in genetic engineering of plants are due to progress in both cellular and molecular biology. The first practical technique for introduction of genetic material into plants was based on the use of vectors from *Agrobacterium tumefaciens*. Due to the narrow host range of this soil bacterium, the scope of the method had been limited to dicotyledonous plants. The progress in cell biology enabling the recovery of intact and fertile plants from single cell and protoplast cultures stimulated research activities in the area of direct DNA transfer. Because it is possible to introduce foreign DNA into protoplasts after electroporation or polyethyleneglycol treatment, the range of plants accessible to genetic engineering was increased significantly. A further decisive breakthrough was achieved by the development of the biolistics approach involving bombardment of the plant tissue with DNA-coated metal particles. It is now possible to engineer almost all important legumes and cereals (1-5).

Genetic engineering offers the possibility not only to add new traits to an organism but also to down-regulate the activities of specific endogenous genes. A versatile and useful method for such a blocking of decisive steps in metabolic pathways is the so-called "antisense technique" (6). The methodology is based on the introduction of an oligonucleotide which is transcribed into messenger-RNA (m-RNA) consisting of sequences complementary to the m-DNA produced through the transcription of the endogenous target gene. Due to the interaction (hybridization) of "sense" and "antisense" RNA, the translation of the corresponding enzyme is strongly reduced. The first genetically modified crop approved for food use, the Flavr-Savr™ tomato, is a prominent example of the commercial exploitation of the "antisense" technology (7).

Analogous to the goals of conventional breeding programs, recombinant DNA techniques are applied to improve (i) agronomic characteristics of crops, such as yield and resistance to diseases and pests, (ii) processing parameters, e.g., optimum solids levels or increased shelf life, and (iii) food quality, including factors such as aroma, taste, and nutritional value. Currently, the majority of commercial applications of plant genetic engineering aim at increasing the yield of food crops by influencing their tolerances to specific herbicides and their resistance to insects and diseases.

Herbicide tolerance. One of the agronomically and commercially important applications, and at the same time an application of genetic engineering strongly opposed by consumer advocates, is the attempt to increase crop yields by increasing the tolerance of plants to particular herbicides. Strategies applied to achieve this goal demonstrate the diversity of tools offered by recombinant DNA techniques: (i) increasing the level of the target enzyme for the herbicide by overexpression, (ii) decreasing the sensitivity of the target enzyme by introducing a gene coding for an enzyme resistant to the herbicide either naturally (from a microbial source) or through specific mutation, and (iii) introducing a gene encoding an enzyme metabolizing and thus detoxifying the herbicide (8, 9). Soybeans tolerant to the herbicide glyphosate, oilseed rape tolerant to phosphinotricin, and cotton exhibiting tolerance to bromoxynil are examples of genetically modified crops presently entering the market (3, 10).

Insect resistance. The use of the bacterium *Bacillus thuringiensis* (*B.t.*) to control plant insect pests has long been known (11). The insecticidal properties are due to the biosynthesis of different toxic crystal proteins. These pro-toxins are proteolytically cleaved in the midgut of the insects, bind to specific membrane receptors and finally lead to the death of the insect. The different strains of *B.t.* produce specific toxins differing in their activity against various classes of insects, such as the larval stages of *Lepidoptera* (moths and butterflies), *Diptera* (flies), and *Coleoptera* (beetles). By means of recombinant DNA techniques it is possible to create plants expressing the *B.t.* δ-endotoxins, thus making them resistant to insect damage. This concept has been applied successfully in commercially important crops, such as cotton, potato, tomato, and corn (3).

Virus resistance. Agronomically important results have been achieved by creating plants resistant to viral infections. The most successful approach, the coat protein mediated protection, is based on the long known observation that infection of a plant with a mild strain of a virus protected it from subsequent infection of a more virulent strain. By means of recombinant DNA techniques it has been possible to express the viral coat protein in plants, thus protecting them against viral diseases. From the first demonstration that the expression of the coat protein gene of the tobacco mosaic virus in tobacco conferred resistance to infection by this virus (12), this approach has been successfully applied in a wide spectrum of plant species (13).

Plants can also be genetically engineered to express proteins which increase their resistance to attacks by fungi or bacteria (14). The increasing understanding of the molecular basis for the phenomenon of "systemic acquired resistance" (15) offers promising applications of recombinant DNA techniques. The strategy involving the expression of bactericidal enzymes from heterologous sources in plants is exemplified by the bacteriophage T4 encoded lysozyme reported to give rise to increased resistance of potatoes to *Erwinia carotovora* (16). Transgenic tobacco expressing a stilbene synthase from peanut was more resistant to infections of *Botrytis cinerea* (17).

Food Quality. Texture, taste, and aroma belong to the major criteria determining the acceptance of a food by consumers. For tomatoes recombinant DNA techniques have proven to be suitable to improve these important attributes. By introducing either the "antisense" (7) or a truncated "sense" (18) polygalacturonase gene, the biosynthesis of this cell wall enzyme, responsible for the breakdown of pectin during ripening, is slowed down. In terms of processing characteristics, this results in optimum solids levels and viscosity. The reduced softening process can also be used either to increase the shelf life of the tomato or to improve its flavor properties by allowing the fruit to remain longer on the vine.

The other approach applied to improve the flavor characteristics of tomatoes is to interfere directly with the metabolism of ethylene, the compound which triggers fruit ripening and aroma formation. By applying the "antisense" technology it is possible to inhibit either the 1-amino-cyclopropane-1-carboxylate (ACC) synthase, the rate-limiting enzyme in the biosynthetic pathway of ethylene (19) or the ACC oxidase, the enzyme involved in the conversion of ACC to ethylene (20).

Potential future strategies to influence flavor properties of a plant have been demonstrated by results of the genetic transformation of a scented *Pelargonium* species, referred to as "lemon geranium". Transformation by means of *Agrobacterium rhizogenes* increased the production of essential oil and significantly changed the distribution of monoterpene alcohols (21).

Nutritional Quality. An area which will attract increasing attention in the near future is the improvement of the nutritional value of foods by means of recombinant DNA techniques. Pioneering examples for changing content and composition of the macronutrients (fats, protein and carbohydrates) have already been described. The direction of fatty acid biosyntheses in favor of medium-chain fatty acids can be achieved by expression of a 12:0-acyl-carrier protein thioesterase in transgenic oilseed plants (22). Increasing the level of sulfur-containing amino acids in soybean by introducing a gene from Brazil nut exemplifies the strategy to improve the balance of essential amino acids in important crops (23). However, this project also demonstrates potential limits, such as the influence of the genetic modification on the allergenicity of the host plant (24). The modification of carbohydrate metabolism by recombinant DNA techniques has been demonstrated for starch in potatoes. The composition of this biopolymer, i.e. the ratio of amylose to amylopectin, as well as its amount in the tubers can be influenced (25,26).

Plants are increasingly being considered as "bioreactors" for production of industrially or pharmacologically important substances, such as proteins (27). Compared to microorganisms the eukaryotic plant cells offer the advantage that the post-translational processing necessary for many valuable proteins can be accomplished. Correctly processed human serum albumin could be obtained from transgenic potatoes (28).

Microorganisms

Microorganisms have played an important role in food production for millennia. The transition from the empirical use of microorganisms to an understanding of the underlying scientific principles was initiated by the pioneering discoveries of Pasteur in the middle of the last century. With the increase of knowledge there have always been attempts to optimize and standardize the microorganisms used in order to meet the requirements of food production. The modern food industry can make use of a spectrum of well-defined "starter-cultures" for the production of fermented foods (29). They are the result of mutagenesis and selection techniques based on classical bacteriological and genetic methods. By means of genetic engineering, the properties of microorganisms can be changed more precisely (30). Major goals are optimization of the production process, improvement of product quality, and safety (hygienic status), and enlargement of product diversity.

Strategies applied are based on (i) increasing the copy number of the gene of interest, (ii) coupling genes with strong promoters or other regulatory sequences, and (iii) expressing genes from other sources in microorganisms which are safe for use in food production (GRAS) and which are easily handled under commercial fermentation conditions. There is a broad array of recombinant microorganisms available for industrial and agricultural applications (31).

Because of its importance both as a model organism in basic research and as a production microorganism in the baking and brewing industry, the yeast *Saccharomyces cerevisiae* has attracted considerable attention among molecular biologists (32, 33). In the United Kingdom genetically modified baker's yeast with increased activities of maltase and maltosepermease (34) and a genetically modified amylolytic brewer's yeast (35) have been reviewed for food use.

Lactic acid bacteria play an outstanding role in food fermentation (36). Accordingly, many examples for their genetic engineering have been reported (37, 38). The construction of safe, so-called "food-grade", vectors has been especially and extensively studied with these microorganisms (39).

Enzymes

For many important processes in food production, enzymes rather than intact microorganisms are employed. These biocatalysts possess outstanding properties, such as substrate specificity, regioselectivity, and enantioselectivity. In particular hydrolases which do not require coenzyme regeneration are increasingly being used commercially. Amylases, pectinases or cellulases are applied in the starch and baking industry as well as in fruit juice production. Lipases are employed to modify the properties of triglycerides by hydrolysis, esterification, and interesterification (40).

Enzymes are isolated from animal, plant, and microbial sources. Pure culture fermentations of selected strains of microorganisms are used to obtain enzyme preparations at industrial scale. Genetic engineering offers the possibility to increase the yield of the desired enzyme by introducing multiple copies of the corresponding gene into the production organism or by influencing the regulatory sequences. A major strategy is to introduce the gene encoding the enzyme in safe and efficient microorganisms. Yeasts have been proven to be ideal expression systems for heterologous proteins (41).

One of the most prominent and pioneering examples for an enzyme obtained from genetically modified microorganisms is the milk-clotting protease, chymosin, the first food ingredient produced via recombinant DNA techniques which has been cleared for food use (42).

In the future, applications of recombinant DNA techniques will not be limited to the production of enzymes, which are structurally and functionally identical to their traditional counterparts. Increasing emphasis will be placed on "protein engineering". This allows for the change of the DNA sequence and subsequently the corresponding amino acid at specific positions, thus designing enzymes with new and optimized properties. Adaptation of enzymes to specific conditions, such as pH and temperature, becomes possible. The use of designed proteases and lipases in the detergent industry indicates the future potential of this strategy in food production.

Food Ingredients

Biotechnological fermentation and biotransformation processes have been used for the production of food ingredients, such as vitamins, amino acids, organic acids, or sweeteners (43, 44). By using the strategies described above for the production of enzymes, recombinant DNA techniques can be applied in order to increase the yield and to improve the recovery and the purification of single compounds from fermentation broths (45).

Because of consumers' demand for "natural" flavor, there has been increasing application of biotechnological methods such as fermentations and biotransformations of corresponding precursors in the production of flavor compounds in recent years (46, 47). Therefore, flavor compounds are ideal examples for food ingredients being suitable for the application of recombinant DNA techniques in the course of the production process.

Animals

The commercial use of recombinant bovine somatotropin (rBST), a growth hormone produced from a genetically modified microorganism, is an example for the *indirect* application of recombinant DNA techniques in farm animal production. In contrast to the progress achieved in genetic engineering of microorganisms and plants, *direct* application resulting in transgenic animals will require more years before it reaches the stage of commercialization. Genetic modification of animals is a much more complex task than that of microorganisms or plants (48). Goals of genetic engineering of animals focus on improved growth and on increased resistance to specific diseases (49). Most of the progress has been made with transgenic fish because the efficiency of integration of DNA into fish is much higher than for mammals (50, 51). Goals such as increased production efficiency and improved growth rate have been achieved by introducing mammalian and fish growth hormone genes. The transfer of the antifreeze protein gene from winter flounder to salmon is an example of improving the tolerance of abiotic stresses.

A promising field is the so-called "gene farming", the production of pharmaceutically or nutritionally important proteins in the mammary glands of transgenic animals (52).

Safety Assessment

The outstanding and fast development of genetic engineering and the accelerating transition from basic research to commercial applications provoke both enthusiastic and very strong negative reactions. Today, it is acceptable to produce life-saving pharmaceuticals by means of genetically modified microoganisms. However, the application of this technology in the production of foods is still the subject of controversial discussion.

The issues being raised can be divided into three major categories: (i) the fact that this technique touches the fundamentals of life by changing the genetic information of organisms provokes ethical concerns; (ii) the release of viable genetically modified organisms into the environment raises ecological issues; (iii) concerns are expressed whether unexpected or unintentional effects will occur as a result of the genetic modification of crops or microorganisms and whether new substances introduced into food will be safe.

National and international organizations are involved in establishing principles for safety evaluations of foods and food ingredients produced via recombinant DNA techniques. The concept of "substantial equivalence" as developed by the OECD (53) is widely agreed upon. It involves the comparison of a food or food ingredient developed by modern biotechnology to its traditional counterpart. If substantial equivalence can be established, the new food or food component can be treated in a similar manner with respect to safety (54).

Current evaluation procedures of genetically modified foods and food ingredients pay particular attention to the safety implications of (i) intentional changes, (ii) any unintentional changes arising from the genetic modification, (iii) the stability of the genetically modified organism under the intended conditions of use; and (iv) the likelihood of genetic transfer.

There is a general consensus that the applicability of classical toxicological assessment procedures developed for single chemical substances, such as pesticides or food additives, is limited. Safety and wholesomeness studies with whole foods have to be carefully designed in order to avoid nutritional imbalances causing artifacts and uninterpretable results.

Particular attention has been paid to the safety evaluation of marker genes and their respective expression products. Marker genes are needed to identify and select cells which have been successfully transformed at an early stage of the genetic modification process. The most important ones are those conferring resistance to antibiotics and tolerance to herbicides, respectively (10). The safety assessment of the antibiotic resistance marker genes has been the subject of detailed investigations (55); international organizations (10), and national regulatory authorities (56) have discussed this issue.

Another concern being raised is whether there is an increased potential for allergenicity of transgenic foods due to the transfer of new proteins. The above mentioned transfer of a gene from Brazil nut to soybean demonstrates that there are methods available to assess the allergenic potential of proteins derived from sources to which consumers have reacted and for which serum is available (24). However, at present the potential allergenicity of proteins that are derived from sources that are not recognized as allergens cannot be predicted.

Regulatory Aspects

There is a relatively broad consensus about the general scientific principles underlying the safety evaluation of foods and food ingredients produced via recombinant DNA techniques. However, the philosophies about regulatory oversight and the need for legislative restrictions vary in different countries. There are two major approaches: one is based on the assumption that the

application of a certain technology, such as genetic engineering, bears the potential
for specific risks and therefore requires a corresponding oversight; the second
approach primarily focusses on the final product and its safety rather than on the
technology applied. These philosophies are reflected in different regulatory
frameworks set up around the world, and some of the world leaders discuss their
regulatory policies in this book.

Literature Cited:

1.   Gasser, C.S; Fraley, R.T. *Science*, **1989**, *244*, 1293-1299.
2.   Gasser, C.S; Fraley. R.T., *Scientific American*, **1992**, *6*, 62-69.
3.   Beck, C.J.; Ulrich, T. *Bio/Technology*, **1993**, *11*, 895-902.
4.   Christou, P. *Agro-Food-Industry-Hi-Tech.*, **1994**, *5*, 17-27.
5.   Willmitzer, L., New Developments in Plant Biotechnology. *Lecture Publications*, Vol. 20, **1994**, Ernst Schering Research Foundation.
6.   Van Blokland, R.; Van der Geest, N,; Mol, J.N.M.; Kooter, J.M. *The Plant Journal*, **1994**, *6*, 861-877.
7.   Sheehy, K.; Kramer, M.; Hiatt, W. *Proc. Natl. Acad. Sci.*, **1988**, *85*, 8805-8809.
8.   Mazur, B.; Falco, S. *Ann. Rev. Plant Physiol.* **1989**, *40*, 441-470.
9.   Oxtoby, L.; Hughes, M. *Trends in Biotechnology* , **1990**, *8*, 61-65.
10. WHO, Health aspects of marker genes in genetically modified plants. *Report of a World Health Organization Workshop*, **1993**.
11. Peferoen, M. *Agro-Food-Industry-Hi-Tech.* **1991**, *2*, 5-9.
12. Powell-Abel, P.; Nelson, R.S; Be, B.; Hoffmann, N.; Rogers, S.G.; Fraley, R.T.; Beachy, R. *Science*, **1986**, *232*, 738-743.
13. Beachy, R.N. In *Biotechnology in Plant Disease Control*, Wiley-Liss, Inc., **1993**, pp. 89-104.
14. Brears, T.; Ryals, J. *Agro-Food-Industry-Hi-Tech.* **1994**, *5*, 10-13.
15. Lawton, K.; Uknes, S.; Friedrich, L.; Gaffney, T.; Alexander, D.; Goodman, R.; Metraux, J.-P.; Kessmann, H.; Ahl, G.; Gut, R.; Ward, E.; Ryals, J. In *Mechanisms of Defense Responses in Plants*, Fritig, B.; Legrand, M., Editors, Kluwer Academic, Dordrecht, **1993**, pp. 410-420.
16. Düring, K.; Porsch, P.; Fladung, M.; Lörz, H. *The Plant Journal*, **1993**, *3*, 587-598.
17. Hain, R.; Reif, H.-J.; Krause, E.; Langebartels, R.; Kindl, H.; Vornam, B.; Wiese, W.; Schmelzer, E.; Schreier, P.H.; Stöcker, R.H.; Stenzel, K.; *Nature*, **1993**, *361*, 153-156.
18. Smith, C. *Mol. Gen. Genet.*, **1990**, *224*, 477-481.
19. Oeller, P.W.; Min-Wong, L.; Taylor, L.P.; Pike, D.A.; Theologis, A. *Science*, **1991**, *254*, 437-439.
20. Hamilton, A.J.; Lycett, G.W.; Grierson, D. *Nature*, **1990**, *346*, 284-287.
21. Pellegrineschi, A.; Damon, J.-P.; Valtorta, N.; Paillard, N.; Tepfer, D. *Bio/Technology*, **1994**, *12*, 64-68.
22. Voelker, T.A.; Worrell, A.C.; Anderson, L.; Bleibaum, J.; Fan, C.; Hawkins, D.J. Radke, S.E.; Davies, H.M. *Science*, **1992**, *257*, 72-74.

23. Townsend, J.A. *Proceedings of the 4th Biennial Conference on Molecular and Cellular Biology of Soybean,* Iowa State University, Ames, IA, **1992**, July 27, p.4.
24. Nordlee, J.A.; Taylor, S.L.; Townsend, J.A.; Thomas, L.A.; Townsend, R. In *OECD Proceedings of a Workshop on Food Safety Evaluation,* Oxford, UK, 12 Sept., **1994**, in press.
25. Müller-Röber, B.; Sonnewald, U.; Willmitzer, L. *EMBO J,* **1992**, *11*, 1229-1238.
26. Stark, D.M.; Timmermann, K.P.; Barry, G.F.; Preiss, J.; Kishore, G.M. *Science,* **1992**, *258*, 287-292.
27. Dale, J.; Belanger, F. *Agro. Food Industry Hi-Tech.* **1993**, *4*, 6-8.
28. Sijmons, P.; Dekker, B.; Schrammeijer, B.; Verwoerd, T.; van den Elzen, P.; Hockema, A. *Biotechnology,* **1990**, *8*, 217-221.
29. *IFBC (International Food Biotechnology Council), Regulat. Toxicol. Pharmacol.* **1990**, *12*, 1-196.
30. Geisen, R; Stander, L.; Leistner, L. *Food Biotechnol.* **1990**, *4*, 497-504.
31. Muraoka, Y.; Imanaka, I., Editors, *Recombinant Microbes for Industrial and Agricultural Applications,* Marcel Dekker, New York, **1993**.
32. Hollenberg, D.P.; Strasser, A.W.M. *Food Biotechnology,* **1990**, *4*, 527-534.
33. Lang-Hinrichs, C.; Hinrichs, J. *Agro. Food Industry Hi-Tech.,* **1992**, *5*, 12-18.
34. *ACNFP (Advisory Committee on Novel Foods and Processes), Report on the use of antibiotic resistance markers in genetically modified food organisms,* **1994**.
35. *ACNFP (Advisory Committee on Novel Foods and Processes), Annual Report,* **1993**, MAFF-Publications, London.
36. Teuber, M. In *Biotechnology, Vol. 1, Biological Fundamentals,* Rehm, H.-J.; Reed, G., Editors, VCH, Weinheim, **1993**, 325-366.
37. De Vos, W.M. In *Harnessing Biotechnology for the 21st Century,* Ladish, M.R.; Bose, A., Editors, American Chemical Society, Washington, DC, **1992**, 524-527.
38. Teuber, M. *Food Reviews International,* **1993**, *9*, 389-401.
39. Venemma, G. *J. Dairy Sci.,* **1993**, *76*, 2133-2144,
40. Graille, J. *Agro-Food-Industry-Hi-Tech.,* **1991**, *6*, 29-35.
41. Gellissen, G. *A. V. Leeuwenhock,* **1992**, *62*, 79-93.
42. Flamm, E.L. *Bio/Technology,* **1991**, *9*, 349-351.
43. Vandamme, E.J., Editor, *Biotechnology of Vitamins, Pigments and Growth Factors,* Elsevier Applied Sciences, London, New York, **1989**.
44. Hodgson, J. *Bio/Technology,* **1994**, *12*, 152-155.
45. Leuchtenberger, A. *Acta Biotechnol.* **1992**, *12*, 57-65.
46. Armstrong, D.A.; Gillies, B.; Yamazaki, H. In *Flavor Chemistry: Trends and Developments,* Teranishi, R.; Buttery, R. G.; Shahidi, F., Editors, ACS Symposium Series 388, **1989**, 105-120.
47. Gatfield, I.L.; Sommer, H. In *Recent Developments in Flavor and Fragrance Chemistry,* Hopp, R.; Mori, K., Editors, VCH, Weinheim, **1993**, 291-304.

48.   Pinkert, C.A., Editor, *Transgenic Animal Technology: A Laboratory Handbook*, Academic Press, San Diego, **1994**.

49.   Brem, G. Transgenic Animals, In *Biotechnology*, Rehm, H.J.; Reed, G.; Puehler, A.; Stadler, P., Editors, VCH-Verlag, **1993**, 745-832.

50.   Chen, T.T.; Lin, C.-M.; Lu, J.K.; Shamblott, M.; Kight, K. In *Science for the Food Industry of the 21st Century, Biotechnology, Supercritical Fluids, Membranes, and other Advanced Technologies for Low Calorie, Healthy Food Alternatives*, Yalpani, M., Editor, ATL Press, **1993**, 145.

51.   *OECD Aquatic Biotechnology and Food Safety*, Paris, **1994**.

52.   Lee, S.H.; de Boer, H.A. *J. Controlled Release*, **1994**, *29*, 213-221.

53.   *Safety Evaluation of Foods Derived by Modern Biotechnology: Concepts and Principles*, OECD, Paris, **1993**.

54.   *Application of the Principles of Substantial Equivalence to the Safety Evaluation of Foods or Food Components from Plants Derived by Modern Biotechnology*, WHO Workshop, WHO, Geneva, in press.

55.   Flavell, R.B.; Dart, E.F; Fuchs, R.L.; Fraley,R.T. *Bio/Technology*, **1992**, *10*, 141-144.

56.   *Report on the Use of Antibiotic Resistance Markers in Genetically Modified Food Organisms*, Advisory Committee on Novel Foods and Processes, (ACNFP), Annual Report, **1993**.

RECEIVED July 14, 1995

REGULATORY OVERSIGHT

Chapter 2

U.S. Food and Drug Administration Policy for Foods Developed by Biotechnology

J. H. Maryanski

Center for Food Safety and Applied Nutrition, U.S. Food and Drug Administration, 200 C Street, S.W., HFS–13, Washington, DC 20204

The Food and Drug Administration (FDA) has authority under the Federal Food, Drug, and Cosmetic Act (the Act) to ensure the safety and wholesomeness of most foods, except meat and poultry, including foods developed through modern biotechnology. In 1990, FDA issued the first regulation for the use of a recombinant DNA-produced food ingredient, fermentation-derived chymosin (rennet). In 1992, FDA published a policy statement that explains how foods and animal feeds derived from new plant varieties developed by both conventional and new breeding techniques are regulated under the Act. The 1992 policy provides "guidance to industry" that establishes a standard of care for ensuring safety and wholesomeness. This discussion summarizes FDA's policy and illustrates how the policy was applied by the agency in reaching decisions on chymosin and on the Flavr Savr tomato.

Foods and food ingredients produced through the techniques of modern molecular biology are now a reality. Over the past four years, the Food and Drug Administration (FDA) has approved the commercial use of chymosin (rennet) produced from bacteria (1,2), yeast (3), and fungi (4) for use in making cheese and other dairy products. In early 1994, FDA determined that the Flavr Savr tomato developed by Calgene, Inc. was as safe as other commercial tomatoes (5). Over 40 food crops modified via recombinant DNA techniques are expected to reach the market in the near future. These crops exhibit improved shelf life, processing characteristics, flavor, nutritional properties, and agronomic characteristics, such as tolerance to chemical herbicides and resistance to pests and disease.

Recombinant DNA techniques are new methods of molecular biology that permit scientists to identify specific genes, make copies of those genes, and introduce the gene copies into recipient organisms, such as a food crop or a microbial starter culture. Once incorporated into the host genome, the introduced gene functions like all other genes in the genome. This process is called transformation, and it is commonly referred to as genetic engineering or gene splicing. Using these techniques, scientists can make copies of genes from any organism--plant, animal, or microbe--from which a potentially useful trait can be identified.

These methods of gene transfer have greatly expanded the pool of potentially useful traits available to scientists for improving food source organisms. Because recombinant DNA techniques are used to introduce one or a few genes into an organism such as a food crop, agricultural scientists avoid one of the major difficulties of conventional cross-hybridization, the concomitant introduction of undesirable genes closely linked to the trait of interest and the subsequent back-crossing necessary to eliminate undesired traits.

The power of genetic modification techniques, in terms of specificity and potentially useful traits, has increased as new methods of gene transfer have been developed (6). Cross-hybridization involves recombination of thousands of genes on whole chromosomes, whereas recombinant DNA techniques are used to transfer or modify one or a few well-characterized genes. Recombinant DNA techniques are used by developers to improve crops as are other methods of genetic modification. These methods can be used as research tools by developers for strain and varietal improvement programs and in conjunction with cross-hybridization, chemical and radiation mutagenesis, somaclonal variation, and embryo rescue.

In spite of the technical advantages of using recombinant DNA techniques, questions have been raised concerning the safety of foods derived using these techniques, especially with respect to the ability to introduce a gene into a food organism from any source. For example, concern is often expressed that new substances whose safety has not been established will be introduced into food, or that unexpected or unintended effects will occur as a result of the newly introduced genetic material, or that new allergens may be present in the food.

FDA's Role in Ensuring Food Safety

The public relies on FDA for assurance that foods are safe and wholesome. FDA has authority under the Act to ensure the safety of most domestic and imported foods in the U.S. market, except meat and poultry, which are regulated by the U.S. Department of Agriculture (USDA). Pesticides used in or on foods are regulated primarily by the Environmental Protection Agency (EPA), which reviews safety and sets tolerances (or establishes exemptions from tolerances) for pesticides. FDA monitors foods to enforce the tolerances for pesticides set by EPA.

FDA regulates foods and food ingredients developed by genetic engineering by the same provisions and regulations under the Act by which it regulates other food products. This means that a food or food ingredient developed by genetic engineering must meet the same rigorous safety standards under the Act as other food products, and FDA has broad authority to take legal action against a substance that poses a hazard to the public.

Chymosin: The First Biotechnology-Derived Food Ingredient

In March 1990, FDA issued the first regulation in the U.S. for the use in food of a substance produced by recombinant DNA techniques (see references *1* and *2*). This substance, chymosin (rennet), is the milk-clotting enzyme used to make cheese and other dairy products. FDA affirmed that chymosin was "generally recognized as safe" (GRAS), meaning that it is exempt from the premarket approval requirements that apply to new food additives. The source of the new enzyme was *Escherichia coli* K-12. Subsequently, chymosin preparations produced from *Kluyveromyces marxianus* var. *lactis* and *Aspergillus niger* var. *awamori* were also affirmed as GRAS (see references *3* and *4*).

FDA considered several important factors in its approval of fermentation-produced chymosin: The introduced chymosin gene encodes a protein that has the same structure and function as animal-derived chymosin; the manufacturing process removes most impurities; the production microorganisms are destroyed or removed during processing and are non-toxigenic and non-pathogenic; and any antibiotic-resistance marker genes (e.g., ampicillin) are destroyed in the manufacturing process.

FDA's Policy for Foods Derived from New Plant Varieties

FDA has, on occasion, been asked questions regarding the safety of new plant varieties, and in the mid-1970s the agency considered criteria by which it would review new varieties developed through conventional breeding (*7*). The scientific community recognized that nutrients and toxicants were important indicators of safety. After the development of molecular biology techniques that could be used to genetically modify food crops in very specific ways, FDA received many questions from developers concerning the safety and regulatory status of these new foods.

In 1992, FDA published and invited public comment on a policy statement (the 1992 policy) clarifying its legal and regulatory framework for oversight of food and animal feed derived from new plant varieties developed by both conventional and new breeding techniques, such as recombinant DNA techniques (*8*). FDA published the 1992 policy to ensure that guidance concerning food safety and regulatory issues was available to developers before products developed by recombinant DNA methods would be ready for safety testing. FDA's policy explains how whole foods, including animal feeds derived from fruits, vegetables, and grains, and by-products such as vegetable oils and food starch, are regulated under the Act. The policy covers foods derived from plants developed through all methods of breeding, including genetic engineering. FDA uses the 1992 policy, on which public comment was requested, as its working policy. It reflects the belief that any policy for a rapidly evolving technology, such as recombinant DNA techniques, should be sufficiently flexible to permit necessary modifications as a result of technological innovations or other information that may come to FDA's attention. For example, FDA is continuing to consider issues raised in comments regarding allergenicity, labeling, and premarket notification.

FDA relies primarily on two sections of the Act to ensure the safety of foods and food ingredients. Generally, whole foods, such as fruits, vegetables, and grains, are not subject to premarket approval. The adulteration provisions of section 402(a)(1) are the primary legal tool that FDA successfully uses to ensure the safety of foods. The Act places a legal duty on developers to ensure that the foods they present to consumers are safe and comply with all legal requirements. FDA has authority to remove a food from the market if it poses a risk to public health. Foods derived from new plant varieties developed through genetic engineering will be regulated under this authority as well.

The FDA also relies on the second section of the Act, the food additive provision (section 409). Under this section, a substance that is intentionally added to food is a food additive, unless the substance is generally recognized as safe (GRAS). Food additives are subject to review and approval by FDA before they may be used in food. Upon request, FDA also reviews and affirms the GRAS status of food ingredients when there is a question regarding the regulatory status of a substance intended for use in food. How this premarket authority will apply to genetic modifications in food crops is discussed below.

The centerpiece of FDA's 1992 policy statement is a comprehensive "guidance to industry" section that discusses scientific issues for ensuring safety and identifies scientific and regulatory questions on which firms should consult with FDA. The agency's guidance to industry establishes a "standard of care" for developers to ensure food safety. The scientific principles that underpin FDA's 1992 policy have been published (9). These principles are consistent with the principles for safety assessment discussed by various prestigious organizations, including the National Research Council in the U.S. (6,10), the World Health Organization and the Food and Agriculture Organization of the United Nations (11), and the Organization for Economic Cooperation and Development (12).

FDA's approach to assessing safety and nutritional composition of a food derived from a new plant variety is predicated on several considerations (13). Today, U.S. grocery stores exhibit a diversity of foods derived from literally hundreds of genetically distinct, new plant varieties, whose safety has been accepted primarily through experience. Rigorous scientific analyses using analytical chemical methods or toxicological studies in animals are rarely conducted. For example, solanine, a glycoalkaloid native to potatoes, is one of the few toxicants in food crops monitored by vegetable breeders in the U.S. Because of the extensive history of safety of plant varieties developed through agricultural research, FDA has not found it necessary to review the safety of foods derived from new plant varieties.

FDA considers the safety of the foods that we have today to be the standard with which the safety of foods derived from new plant varieties should be compared (see references 11-13). The safety assessment approach outlined in FDA's 1992 policy focuses on the intended genetic modification and the overall composition of important nutrients and toxicants in the food. This concept recognizes that although new foods are variants of existing, well-accepted foods, these well-accepted foods are not inherently safe. That is, many foods contain components that would present safety concerns if those substances were present in the food in concentrations above the range that has been found to be acceptable. In addition, some individuals in the population are allergic or intolerant to certain foods. Thus, a level of absolute safety for a food cannot be achieved or expected.

Developers should evaluate a food derived from a new plant variety by comparison with other commercial varieties of the crop. This approach is widely accepted in the scientific community. It is also recognized that foods--fruits, vegetables, and grains--consist of complex mixtures of many substances. The accepted approach for assessing the safety of foods differs from approaches applied to single chemical substances such as food additives and pesticides, in which safety is generally established by non-clinical studies in animals. Animal feeding studies with foods are usually not sufficiently sensitive to detect toxic constituents in the food, and it is usually not possible to supplement the diet with a high enough concentration of test material to achieve the desired safety margin. In addition, high concentrations of food added to the diet can perturb the nutritional balance of the diet and confound interpretation of the results.

To circumvent the difficulties of tests in animals, a multidisciplinary approach is used to evaluate the safety and nutritional composition of a food. This approach relies on information pertaining to the agronomic and quality attributes of the plant, genetic analysis of the modification and stability of expected genomic traits (e.g., Southern analysis of the introduced gene(s) and restriction fragment-length polymorphisms), evaluation of the safety (toxicity and allergenicity) of newly introduced proteins, and chemical analyses for important toxicants and nutrients. If safety questions remain after this evaluation, toxicological studies can be designed to address them.

The guidance to industry section of the 1992 policy focuses on issues related to changes in food crops that are both intended and unintended or unexpected modifications of the finished food. FDA begins with the premise that many varieties of food crops have been developed through plant breeding and that the foods derived from these varieties are generally safe for consumption, although there have been rare exceptions. FDA's guidance addresses safety issues relevant to the food crop that is being modified, the potential for any introduced genetic material to encode harmful substances, the safety of intentionally introduced substances (e.g., proteins encoded by introduced genes), and the assessment of acceptable levels of known plant toxicants and important nutrients in the new variety. This guidance is presented in a series of flow charts and text that covers the food crop being modified, the source(s) of any introduced genetic material, and new substances intentionally added to the food as a result of the genetic modification, i.e., proteins, fatty acids, and carbohydrates.

One important feature of the 1992 policy is FDA's requirement for premarket approval, as food additives, of proteins (or other added substances such as fatty acids and carbohydrates) produced by introduced genes if the protein differs substantially in structure and function from the many proteins that comprise our foods. Conversely, FDA presumes that proteins which are derived from foods and proteins which are substantially similar to those that comprise our foods are GRAS. In these cases, premarket review is generally not required. From its present knowledge of developments in agricultural research, FDA believes that most of the substances that are being introduced into food by genetic modification have been safely consumed as food or are substantially similar to such substances. Therefore, FDA does not anticipate that most foods developed by recombinant DNA methods will contain substances that require premarket approval as new food additives.

One issue related to the transfer of genetic material between organisms that FDA believes deserves particular attention is the possibility that proteins which have been introduced into a food could cause allergic reactions in some individuals. We believe that particular attention should be given to proteins that are derived from foods to which individuals in the U.S. population are commonly allergic, such as milk, eggs, wheat, fish, tree nuts, and legumes. In such cases, the developer should demonstrate scientifically that the allergenic substance is not present in the new food, or FDA should require some form of labeling to alert sensitive consumers.

In April 1994, FDA, EPA, and USDA hosted a scientific conference on "Scientific Issues Related to Potential Allergenicity in Transgenic Food Crops" (*14*, transcript available from FDA as Docket No. 94N-0053). The goal of the conference was to assess current information regarding the attributes of substances (such as proteins) that are food allergens by fostering a dialogue among scientists on food allergy and new varieties of food crops developed by gene transfer. The scientists presented and discussed papers on plant breeding and biotechnology, allergenic foods, exposure and allergic response, T cell and B cell antigenic determinants, *in vitro* and *in vivo* diagnostics, and animal models. They noted that allergic reactions to foods occur in a small percentage of the U.S. population but, nevertheless, affect a significant number of individuals. Life-threatening reactions are a rare occurrence, and most allergic reactions to foods can be attributed to fewer than a dozen foods. Methods are available to assess allergenic potential for proteins that are derived from sources to which consumers have reacted and for which serum is available, but it may be useful to establish a serum bank. There are no direct methods to assess potential allergenicity of proteins from sources that are not known to produce food allergy. Although the possibility that a new protein will cause an allergic reaction can be minimized by evaluating its similarity to known food allergens (i.e., whether the new protein has a molecular size and an amino acid sequence similar to those of known allergens and whether the new protein is resistant to degradation by heat, acid, and gastric enzymes), no one factor is predictive. Glycosylation of the protein was not considered a useful parameter.

The goal of a safety and nutritional assessment should be to establish that the new food is as safe as the foods in U.S. grocery stores today. As we have said previously (see reference *9*), "FDA's science-based approach for ensuring the safety of foods from new plant varieties focuses safety evaluation on the objective characteristics of the food: The safety of any newly introduced substances and any unintended increased concentrations of toxicants beyond the range known to be safe in food or alterations of important nutrients that may occur as a result of genetic modification. Substances that have a safe history of use in food and substances that are substantially similar to such substances generally would not require extensive premarket safety testing. Substances that raise safety concerns would be subjected to closer inquiry. This approach is both scientifically and legally sound and should be adequate to fully protect public health while not inhibiting innovation."

Evaluation of the Flavr Savr Tomato

The first food derived from a crop modified via recombinant DNA techniques to come before FDA was the Flavr Savr tomato developed by Calgene, Inc. (Calgene) of Davis, California (*15,16*; see also FDA Docket No. 91A-0330). To develop this tomato, Calgene used recombinant DNA techniques to introduce an antisense polygalacturonase (PG) gene into the tomato. The sense PG gene, normally present in tomatoes, encodes the enzyme PG, which is associated with the breakdown of pectin (a constituent of the tomato cell wall) and the resulting softening of ripe tomatoes. The antisense PG encodes a messenger RNA that suppresses the production of the PG enzyme. The result is a tomato that remains on the vine longer for enhanced flavor.

In developing the Flavr Savr tomato, Calgene used a selectable marker gene, kanamycin resistance, that encodes the enzyme aminoglycoside-3'-phosphotransferase II (APH(3')II) to identify plant cells carrying the antisense PG gene. APH(3')II inactivates the antibiotics kanamycin and neomycin, and its presence in plant cells permits cells to survive and grow in the presence of these antibiotics, unlike normal plant cells, which are killed by these antibiotics. This allows scientists to select transformed cells that have successfully taken up the desired PG gene.

Calgene asked FDA to evaluate the Flavr Savr tomato under the most stringent procedures available for foods to ensure public confidence in its product. Thus, in addition to evaluation of the firm's safety and nutritional assessment of the tomato *per se*, Calgene requested that FDA regulate the APH(3')II enzyme, the only new substance in the Flavr Savr tomato, as a food additive (for details see FDA Docket Nos. 90A-0416 and 91A-0330).

Overall, FDA evaluated the data and information provided by Calgene to determine whether Flavr Savr tomatoes were significantly altered, compared with varieties of tomatoes with a safe history of use. In other words, FDA asked, "Are Flavr Savr tomatoes as safe as other currently consumed tomatoes?"

On the basis of the safety and nutritional assessment described in its 1992 policy and the modifications of the Flavr Savr tomato, FDA believes that this new tomato should be evaluated by an analysis of the following information: the source, identity, function, and stability of genetic material introduced into Flavr Savr tomatoes; analytical studies on the composition of Flavr Savr tomatoes; and the safety of APH(3')II. FDA also evaluated the environmental safety of the use of the kanamycin resistance gene as part of its review of the food additive petition for APH(3')II.

The DNA introduced into the Flavr Savr tomato was derived from *Agrobacterium tumefaciens, E. coli*, cauliflower mosaic virus, and the tomato. Calgene demonstrated that APH(3')II was the only full-length gene encoded by the introduced genetic material. The firm also showed that the introduced DNA was stably integrated in the tomato chromosome and remained unchanged over five generations.

Calgene compared the nutritional profile of Flavr Savr tomatoes with the parental variety to ensure that the new tomato did not exhibit unexpected changes in composition. Due to the high consumption of tomatoes and tomato products in the United States, tomatoes are an important source of vitamins A and C. Calgene analyzed representative fruits for these vitamins during storage under conditions expected for commercial tomatoes. The firm found no significant difference between the Flavr Savr tomato lines and the control parental line. Calgene also found no difference between the Flavr Savr tomato lines and the control parental line in lycopene or beta-carotene content.

Discussions with plant breeders have indicated that developers do not routinely analyze new tomato varieties for the naturally occurring glycoalkaloid tomatine. However, Calgene wished to provide assurance to its consumers that unexpectedly high levels of this toxicant do not occur in Flavr Savr tomatoes. Tomatine is known to occur in mature green tomato fruit, but tomatine concentrations decrease as the fruit ripens. Calgene showed that there were no significant differences between the glycoalkaloid content of Flavr Savr tomatoes and those of commercial tomato varieties at both mature green and red-ripe stages of development.

The only new substance introduced into the Flavr Savr tomato was the APH(3')II marker gene protein. General considerations for the safe use of marker genes have been established (*17*). Calgene evaluated the safety of this protein (*18*) and showed that APH(3')II is rapidly inactivated by stomach acid and digestive enzymes. The firm noted that enzymes such as APH(3')II are heat labile. The enzyme also is not significantly homologous with any proteins listed as food allergens or toxins. Furthermore, APH(3')II is a phosphorylating enzyme, a type of enzyme commonly found in edible plants and animals. Finally, the enzyme occurs in food at very low concentrations (conservatively estimated at 0.16 parts per million in the diet, based on a 100% market share for tomatoes containing APH(3')II). FDA concluded that APH(3')II does not possess any of the recognized characteristics of food allergens or any attributes that would distinguish it toxicologically from other phosphorylating enzymes in food.

Calgene also considered whether APH(3')II could affect the therapeutic efficacy of orally administered aminoglycoside antibiotics. Even though the enzyme had shown rapid degradation under normal gastric conditions, Calgene evaluated whether a significant amount of orally administered antibiotic could be inactivated by APH(3')II under abnormal stomach conditions, such as may exist in patients treated with drugs that reduce stomach acidity, in which the enzyme might survive digestion. Because APH(3')II requires the cofactor ATP for enzyme activity, Calgene considered whether the amount of ATP available in food would be sufficient to result in the inactivation of a significant amount of orally administered antibiotic. Calgene's worst-case assessment (high intake of ATP-containing food, low dose of antibiotic) showed that only a small fraction of the antibiotic would be inactivated. The firm also showed that no significant inactivation of kanamycin was observed during *in vitro* studies on tomato extract containing APH(3')II and kanamycin.

Calgene also considered whether the kanamycin resistance gene present in the Flavr Savr tomato chromosome could be transferred to pathogenic microbes in the intestinal tract or in soil, rendering the microbes refractory to the effects of the antibiotic. There is no known mechanism by which a gene can be transferred from a plant chromosome to a microbe. Thus, the possibility that such transfer would generate new resistant organisms is very small, especially when compared with the high rate of spread of resistance through known mechanisms of microbe-to-microbe transfer of antibiotic resistance genes.

Based on the information that Calgene submitted concerning the Flavr Savr tomato, FDA concluded that this new variety had not been significantly altered in regard to safety when compared with varieties of tomatoes with a safe history of use. FDA also concluded that the only new substance in the tomato, APH(3')II, was safe for consumption when present in tomatoes at the concentrations typically found in food derived from plants transformed by the use of this selectable marker.

Labeling

FDA's May 1992 policy addressed the labeling of foods derived from new plant varieties, including plants developed by genetic engineering (see reference 8, p 22991). The Act defines the information that must be disclosed in labeling (including information on the food label) and requires that all labeling be truthful and not misleading. The Act does not require disclosure in labeling of information solely on the basis of consumers' desire to know. The Act does require that a food be given a common or usual name, and that the label disclose information that is material to representations made or suggested about the product and consequences that may arise from the use of the product.

FDA will require special labeling if the composition of a food developed through genetic engineering or any other method differs significantly from its conventional counterpart. For example, if a food contains a major new sweetener as a result of genetic modification, a new common or usual name or other labeling may be required. Similarly, if a new food contains a protein derived from a food that commonly causes allergic reactions (and the developer cannot demonstrate that the protein is not an allergen), labeling would be necessary to alert sensitive consumers because they would not expect to be allergic to that food. However, if a protein commonly produces very serious allergic reactions (e.g., peanut protein) and is transferred to another food, FDA would need to evaluate whether it would be practicable to label the food throughout its distribution. Circumstances could exist for which labeling would not provide sufficient consumer protection, and FDA would take appropriate steps to ensure that the food would not be marketed.

To date, FDA is not aware of information that would distinguish genetically engineered foods as a class from foods developed through other methods of plant breeding and, thus, require such foods to be specially labeled to disclose the method of development. The agency has not required labeling for other methods of plant breeding such as chemical- or radiation-induced mutagenesis, somaclonal variation, or cell culture. For example, sweet corn is not required to be labeled "hybrid sweet corn" because it was developed through cross-hybridization.

FDA did not require special labeling for the Flavr Savr tomato. The agency concluded that the correct common or usual name for the Flavr Savr tomato is "tomato" because it is not significantly different from the range of commercial varieties referred to by that name. However, Calgene has decided to provide special labeling, including point-of-sale information, to inform consumers that the new tomato has been developed through genetic engineering (see reference 5).

Summary

Irrespective of the method by which a food or food ingredient is produced, all products must meet the same stringent safety standards and be properly labeled in accordance with the Act. Our approval of fermentation-produced chymosin illustrated an approach for assessing safety of substances derived from genetically modified sources. FDA has since provided guidance for developers that establishes a standard of care to ensure that foods derived from new plant varieties are safe and wholesome. FDA evaluated the data and information supplied by Calgene and agreed with the firm that the Flavr Savr tomato is as safe as other commonly consumed tomatoes.

Postscript

Since its decision on the Flavr Savr tomato, FDA has asked developers of foods derived from new plant varieties developed by using recombinant DNA techniques to provide only summary information of their safety and nutritional assessment to FDA and to make a scientific presentation of their data to FDA scientists. This informal notification process serves to inform the agency about developments in the technology and permits FDA to identify any unresolved safety or regulatory questions. In November 1994, FDA completed informal notifications with developers on seven additional foods derived from plants modified via recombinant DNA techniques, and presented the safety and nutritional summary information on the products to the agency's Food Advisory Committee. These foods included delayed-ripening tomatoes (DNA Plant Technology, Monsanto Co. and Zeneca Plant Sciences); pest-resistant crops: virus-resistant squash (Asgrow Seed Co.) and Colorado potato beetle-resistant potato (Monsanto Co.); herbicide-tolerant crops: bromoxynil-tolerant cotton (Calgene, Inc.), and glyphosate-tolerant soybean (Monsanto Co.). The Food Advisory Committee agreed that there are no outstanding food safety issues associated with these products.

Acknowledgments

The author gratefully acknowledges the guidance and friendship of Richard J. Ronk, who retired from FDA in 1994 after over 40 years of government service. His advice was invaluable in the formulation of FDA's food biotechnology policy. The author also appreciates the critical review of the manuscript by Laura Tarantino, FDA, Washington, DC.

Literature Cited

1.   U.S. Food and Drug Administration. Direct food substance affirmed as generally recognized as safe; chymosin enzyme preparation derived from *Escherichia coli* K-12, *Fed. Regist*. **March 23, 1990**, *57*, 10932-10936.
2.   Flamm, E.L. *Bio/Technology* **1991**, *9*, 349-351.
3.   U.S. Food and Drug Administration. Direct food substances affirmed as generally recognized as safe; chymosin enzyme preparation derived from genetically modified *Kluyveromyces marxianus* (Hansen) Van Der Walt variety *lactis* (Dombrowski) Johannsen et Van Der Walt, *Fed. Regist*. **February 25, 1992**, *57*, 6476-6479.

4. U.S. Food and Drug Administration. Direct food substances affirmed as generally recognized as safe; chymosin enzyme preparation derived from *Aspergillus niger* Van Tieghem variety *awamori* (Nakazawa) Al-Musaliam, *Fed. Regist.* **May 7, 1993**, *58*, 27197-27203.

5. Kahl, L.S. Summary of Consultation with Calgene, Inc., Concerning Flavr Savr Tomatoes, FDA Docket No. 91A-0330, Memorandum, HFS-206 to HFS-200; FDA: Rockville, MD, May 17, 1994.

6. National Research Council. *Field Testing Genetically Modified Organisms: Framework for Decisions*; National Academy Press: Washington, DC, 1989; p 16.

7. Spiher, A.T. *HortScience* **1975**, *10*, 241-242.

8. U.S. Food and Drug Administration. Statement of policy: foods derived from new plant varieties, *Fed. Regist.* **May 29, 1992**, *57*, 22984-23005.

9. Kessler, D.A.; Taylor, M.R.; Maryanski, J.H.; Flamm, E.L.; Kahl, L.S. *Science* **1992**, *256*, 1747-1832.

10. National Research Council. Introduction of Recombinant DNA-Engineered Organisms into the Environment: Key Issues; National Academy Press: Washington, DC, 1987.

11. Report of a Joint FAO/WHO Consultation. *Strategies for Assessing the Safety of Foods Produced by Biotechnology*; World Health Organization: Geneva, 1991.

12. Organization for Economic Cooperation and Development (OECD). *Safety Evaluation of Foods Derived by Modern Biotechnology*; OECD: Paris, 1993.

13. Ronk, R.J.; Maryanski, J.H.; Thompson, P. *HortScience* **1990**, *25*, 1482-1484.

14. U.S. Food and Drug Administration. Conference on scientific issues related to potential allergenicity in transgenic food crops, *Fed. Regist.* **April 1, 1994**, *59*, 15415.

15. Redenbaugh, K.; Hiatt, W.; Martineau, B.; Kramer, M.; Sheehy, R.; Sanders, R.; Houck, C.; Emlay, D. *Safety Assessment of Genetically Engineered Fruits and Vegetables*, CRC Press, Inc.: Boca Raton, FL, 1992.

16. Redenbaugh, K.; Berner, T.; Emlay, D.; Frankos, B.; Haitt, W.; Houck, C.; Kramer, M.; Malyj, L.; Martineau, B.; Rachman, N.; Rudenko, L.; Sanders, R.; Sheey, R.; Wixtrom, R. *In Vitro Cell. Dev. Biol.* **1993**, *29P*, 17-26.

17. World Health Organization. Health Aspects of Marker Genes in Genetically Modified Plants: Report of a WHO Workshop; WHO: Geneva, 1969.

18. U.S. Food and Drug Administration. Secondary direct food additives permitted in food for human consumption; food additives permitted in feed and drinking water of animals; aminoglycoside 3'-phosphotransferase II, *Fed. Regist.* **May 23, 1994**, *59*, 26700-26711.

RECEIVED June 7, 1995

Chapter 3

Regulatory Oversight and Safety Assessment of Genetically Modified Foods in the European Union

K.-H. Engel, M. Schauzu, G. Klein, and A. Somogyi

Federal Institute for Health Protection of Consumers and Veterinary Medicine, Postfach 330013, D–14191 Berlin, Germany

The regulatory bodies of the European Union (EU) are currently preparing a legislative framework for the placing on the market of foods resulting from new technologies, processes, and materials, including genetic modification. The Commission has presented a proposal for a "Regulation on Novel Foods and Novel Food Ingredients" which is presently going through the EU legislative process. The underlying principle of this regulation is a premarket human and environmental safety assessment and, if necessary, an authorization decision prior to marketing. The incorporation of this approach in the EU concept of regulatory oversight on the application of biotechnology/genetic engineering is presented. The guidelines being developed for the safety assessment of genetically modified foods are outlined.

The advances in genetic engineering and the increasing application of this technology in many areas gave rise to the development of a common regulatory framework for biotechnology in the European Union (EU). The intention of this legislation is both, to ensure protection of human, animal and plant health and the environment from hazards from the deliberate release of genetically modified organisms (GMOs) into the environment and to encourage research, development, and commercialization of modern biotechnology.

The functioning of a common market without internal frontiers and a free movement of goods, persons, services and capital requires an approximation of the laws of the Member States. One of the areas where such a harmonization of legislative requirements and authorization procedures is needed is the production of foods and food ingredients from genetically modified organisms. The legislation on genetically modified foods in the EU will be incorporated in an overall regulatory framework for biotechnology. The rationale and the major elements of legislative requirements and the principles for the safety assessment of genetically modified foods will be summarized.

The European Union Regulatory Framework for Biotechnology
There are two different regulatory philosophies determining the creation of a legislative framework for biotechnology:
• *a technology related* approach, based on the assumption that the application of a certain technology may cause specific risks and requires a corresponding oversight;
• *a product related* approach, which primarily focusses on the final product and its safety rather than on the technology applied.
The EU regulatory framework for biotechnology will be a combination of both approaches. The technology related approach is reflected in the "**horizontal**" **legislation** covering all stages of the application of biotechnology in order to ensure the protection of environment and human health from *technology related* risks. The planned "**vertical**" **legislation** will relate specifically to characteristics of sectors and products affected by biotechnology in order to ensure that testing and authorization procedures are streamlined and that one assessment and notification procedure covers all requirements for product authorization.

"Horizontal" Legislation
In adopting the EU legislation on the application of modern biotechnology, it was recognized that the release of organisms having a combination of traits that nature may have never produced increases uncertainty as regards the behavior of the organisms and the possiblity of an adverse impact on the environment. The legislation therefore foresees that an environmental risk assessment must always be carried out before any release of GMOs into the environment whether for an experiment or in a product and that no release may be carried out without the consent of the competent authorities.
The biotechnology regulatory framework is aiming at a preventive approach and is therefore based on the "step by step" principle in order to guarantee that unknown risks are assessed at an early stage. This means that the containment of GMOs is reduced and the scale of release is increased gradually, but only if evaluation of the earlier steps in terms of protection of human health and the environment indicates that the next step can be taken.
The "horizontal" legislation framework comprises three directives:
• Council Directive of 23 April 1990 on the Contained Use of Genetically Modified Microorganisms (*90/219/EEC*) (1). According to this directive appropriate containment measures have to be applied at the various operations involving genetically modified microorganisms in order to control emissions and to prevent accidents. Genetically modified microorganisms are classified in relation to the risks they present. Depending on the degree of risk involved, a notification of the contained use of genetically modified microorganisms or a consent of the national competent authority is required.
• Council Directive of 26 November 1990 on the Protection of Workers From Risks Related to Exposure of Biological Agents at Work (*90/679/EEC*) (2). In this directive protective measures have been set up in order to protect the health and safety of workers exposed to biological agents, including genetically modified microorganisms.
• Council Directive of 23 April 1990 on the Deliberate Release into the

Environment of Genetically Modified Organisms (*90/220/EEC*) (3). This directive foresees that a case-by-case environmental risk assessment has always to be carried out prior to a release of GMOs into the environment. The legislation covers the research stage, where in most cases the deliberate release of GMOs is a necessary step in the development of a product, as well as the placing on the market of the product. According to the "step by step" principle no product containing, or consisting of, GMOs and intended for deliberate release will be considered for placing on the market without it first having been subject to satisfactory field testing at the research and development stage in ecosystems which could be affected by its use.

Before consent according to Directive *90/220/EEC* can be given to a GMO containing product, the manufacturer or the importer to the Community has to demonstrate on the basis of the results of previous releases notified under this Directive, or on substantive, reasoned scientific grounds, that commercialization and use of a GMO containing product do not pose a risk to human health and the environment.

The following authorization procedure has been established for the placing on the market of products containing, or consisting of, GMOs:

• Notifications have to be submitted to the competent authority of the Member State in which the product will be marketed for the first time. This authority will act as a "gateway" to the system and will be responsible for carrying out the main environmental risk assessment and for issuing the final consent. The applicant has to submit with his application:

(i) information on data and results obtained from research and development releases concerning the ecosystems which could be affected by the use of the product and an assessment of any risks for human health and the environment related to the GMOs contained in the product;

(ii) the conditions for the placing on the market of the product, including specific conditions of use and handling and a proposal for labeling and packaging.

• The competent authority must respond within 90 days by either forwarding the application dossier to the Commission with a favorable opinion for consent, or informing the applicant that the application is to be rejected.

• If the dossier is forwarded to the Commission, the Commission is responsible for ensuring that a Union-wide consultation procedure takes place. The competent authorities of the Member states have 60 days within which to register objections.

• If there are no objections, written consent valid for the whole Union is given by the competent authority which received the notification. If there are objections, a specific Commission decision is required.

As long as specific product related legislations are not in force, Directive *90/220/EEC* covers the commercialization of all products containing or consisting of GMOs, including foods and food ingredients. The term "organism" within the meaning of this directive is defined as biological entity capable of replication or of transferring genetic material. Therefore, the placing on the market of genetically modified tomatoes or yoghurt containing viable genetically modified lactic acid bacteria would be examples to fall in the scope of Directive

90/220/EEC; however, products, such as paste made from genetically modified tomatoes or the milk-clotting enzyme chymosin isolated from genetically modified microorganisms, are not covered by this directive.

"Vertical" Legislation

In order to avoid multiple authorization steps, to concentrate administrative resources, and to reduce excessive paperwork the EU regulatory framework for biotechnology will include a product related "vertical" legislation. According to the so-called *"one door - one key"* principle, this concept is based on the idea to provide a single, integrated notification and assessment procedure for placing products developed by biotechnology on the market. Therefore, the provisions of Directive *90/220/EEC* relating to the placing on the market of products containing, or consisting of, GMOs do not apply once a product related legislation considering a specific environmental risk assessment similar to that laid down in this Directive is in force. The Directives on the placing on the market of plant protection products (4) and of medicinal products (5), the currently discussed amendment of the regulations on seeds and the proposal for a European Regulation on Novel Foods and Novel Food Ingredients are examples for this philosophy.

Proposal for a European Regulation on Novel Foods and Novel Food Ingredients

At present there is no general requirement in the EU Member States that products offered for sale as foods are subjected to a premarketing assessment. Companies or persons can place foods on the market on their own responsibility. However, authorities have the power to take action if they consider that a food is a danger for health.

The European Parliament and the Council of the European Union are currently discussing provisions at Union level for the placing on the market of foods resulting from a new range of materials, processes and technologies, including genetic modification, which are summarized under the term "Novel Foods".

A first proposal for a "Council Regulation on Novel Foods and Novel Food Ingredients" has been presented by the Commission of the European Communities in July 1992 (6). This proposal has been going through the Community legislative process. In October 1993 the European Parliament adopted its opinion on the proposal. The Commission took account of the Parliament's opinion and presented an amended proposal in December 1993 (7).

The regulation will provide a scheme for those responsible for placing foods on the market and also for the control authorities to identify those cases where there is a need to scientifically evaluate a food which is being offered for sale for the first time. The underlying principle is a premarket safety assessment of novel foods and novel food ingredients and, if necessary, an authorization decision prior to marketing with different responsibilities being assigned to the Member States and the Commission. There is a procedure for objections, leading to either a "green light" for the applicant to proceed with marketing the food, or to a formal authorization decision.

Scope of the Regulation

Novel foods and novel food ingredients within the meaning of the Regulation are those which have not hitherto been used for human consumption to a significant degree and which fall under the following categories:

• foods and food ingredients, containing or consisting of GMOs within the meaning of Directive *90/220/EEC*;

• foods and food ingredients produced from GMOs, except for those foods produced from GMOs which, by comparison with conventionally manufactured products, have not undergone any significant change in their composition, nutritional value or intended use;

• food and food ingredients with a new or intentionally modified primary molecular structure which have not normally been used hitherto as food or food ingredients;

• single-cell proteins intended for food production;

• foods and food ingredients to which has been applied a process not currently used in food production or which, although subjected to such a process have not previously been placed on the market and where such a process gives rise to significant changes in composition or structure of the end product which affect its nutritional value, digestibility, metabolism or level of undesirable substances in food.

The Commission has asked its advisory body, the Scientific Committee for Food (SCF), to develop explanatory guidelines concerning the categories of novel foods and food ingredients falling in the scope of the regulation. The Regulation will not apply to food additives, flavorings for use in foods, extraction solvents used in the production of foods, and foods and food ingredients treated with ionizing radiation, for which other Union provisions are applicable.

There is a general consensus among the Member States that foods and food ingredients containing or consisting of GMOs within the meaning of Directive *90/220/EEC* should fall in the scope of the Regulation. Discussions are still continuing on the question whether to exempt foods and food ingredients produced from GMOs but which are "substantially equivalent" to traditional counterparts from the scope.

Criteria for Authorization

Before a novel food product is released onto the market of the European Union (EU) the producer will have to demonstrate that the product complies with the following criteria:

• The products are safe for the consumer when consumed at the intended level of use.

• The products do not mislead the consumer.

• The products do not differ from similar foods or food ingredients that they may replace in such a way that their normal consumption would be nutritionally disadvantageous for the consumer.

Initial Assessment

• The applicant has to submit a request to the Commisssion and the Member States comprising the necessary information, including a copy of the studies which have been carried out and all the other evidence which is available to demonstrate that the food or food ingredient complies with the above mentioned criteria.

• Each Member State appoints a competent food assessment body. The Commission will arrange for one of the competent food assessment bodies of the Member States to prepare and transmit an initial assessment report to the Commission within a period of two months. The Commission will circulate the initial assessment report to all Member States.

• Any Member State or the Commission may, within a period of thirty days, make comments or present reasoned objections to the marketing of the product concerned.

• In the absence of objections and if the product is not consumed as a viable organism the Commission will inform the applicant that he may proceed with the placing on the market.

Authorization Procedure

• In the case of objections or when the product contains or consists of GMOs within the meaning of Directive *90/220/EEC*, a decision on the authorization for the marketing of the product is required by the Commission. In the case of GMO containing products the decision will respect the environmental safety requirements laid down by Directive *90/220/EEC*.

• The procedure for the adoption of an authorization decision foresees the assistance of the Standing Committee of Foodstuffs (Committee) which consists of representatives of the Member States and is chaired by a representative of the Commission.

The Commission will consult the SCF and will take the advice of the Committee into consideration preceding a decision or provision regarding a food or food ingredient likely to have an effect on public health, either on its own initiative or at the request of a Member State.

Where a Member State has detailed grounds for considering that the use of a food or food ingredient, although it complies with this regulation, endangers human health, the Member State may temporarily suspend or restrict the trade and use of the product in question in its territory. The Commission will examine the grounds given by the Member State within the Committee and the SCF, and will adopt the necessary measures.

Labeling Provisions

The decision on authorization will establish the conditions of use and the name of the food or food ingredient as well as whether and to what extent requirements for labeling have to be made.

In addition to the general requirements laid down in other regulations or directives relating to the labeling, presentation and advertising of foods for sale to the ultimate consumer, specific labeling provisions may be required. They should ensure the information of the consumer on any significant differences in the characteristics of the novel food or food ingredient when compared with the equivalent conventional counterpart.

Due to the importance of labeling provisions in public discussions and the expected impact on the acceptance of novel foods and novel food ingredients by the consumers, this aspect is one of the key issues being raised in the present discussion of the amended proposal in the legislative bodies of the EU.

Once the Council and the Parliament have adopted the Regulation on Novel Foods and Novel Food Ingredients it will be binding in its entirety and directly applicable in all EU Member States.

Strategies for Assessing the Safety of Genetically Modified Foods

International Organizations

Representatives from the Member States of the European Union have been participating in activities of national and international organizations to develop criteria and guidelines for the safety assessment of foods and food ingredients produced via recombinant DNA techniques.

The objective of a Joint FAO/WHO Consultation held in Geneva, 1990, was to reach international consensus on broad strategies for assessing the safety of foods produced by biotechnology (8). As a continuation of that work a WHO Workshop hosted by the National Food Agency of Denmark addressed the more specific and detailed issues related to the use of marker genes in genetically modified plants (9).

A concept of basing safety evaluations of a new food or food component on a comparison to an existing traditional counterpart has been developed by the OECD Group of National Experts on Safety in Biotechnology (10). Practical guidance on how to apply this principle of "substantial equivalence" to foods or food components from plants derived from modern biotechnology has been elaborated on a recent WHO-Workshop held in Copenhagen (11).

The question of what strategies can be used to establish the safety of foods produced by biotechnology if there is no acceptable counterpart for comparison has been addressed on an OECD Workshop in Oxford (12). The demand for increased knowledge in the safety evaluation of novel foods has been the subject of a symposium sponsored by the German Ministry of Research and Technology at the Hohenheim University (13).

Despite the different philosophies regarding the need of regulatory oversight and authorization procedures there is a general consensus on the major scientific principles underlying the safety assessment of foods produced via recombinant DNA techniques. Key steps in a multidisciplinary approach are:

• the characterization of the genetic modification,
• the consideration of the agronomic performance (for plants),
• the evaluation of the toxicity of any newly introduced compound and of the allergenicity of new proteins,
• the compositional analysis of critical toxicants and nutrients.

If on the basis of these data "substantial equivalence" of a new food or food component to an existing counterpart can be established, it can be treated in the same manner with respect to safety (10). If "substantial equivalence" to a counterpart with an accepted standard of safety cannot be found and any safety questions remain, carefully designed toxicological studies may be conducted in order to specifically address these issues.

Scientific Advisory Committees

Several European States have established scientific committees of independent experts to give advice to their governments and guidance to the industry on the safety assessment of novel foods and processes. The Scandinavian countries have set up the Nordic Working Group on Food Toxicology and Risk Assessment in 1988. This advisory body has published a report in 1991 on "Food and new biotechnology - novelty, safety and control aspects of foods made by new biotechnology" (14).

The Netherlands Food and Nutrition Council has issued an "Advisory Report on Biotechnology" in 1993, which indicates (i) which aspects of the application of new biological techniques in the production of foods require particular attention of the government and (ii) how the consumers' need for information on the application of these techniques in the production of foods can be satisfied as responsibly and effectively as possible (15).

The report points out that not only the traditional toxicological aspects, but especially the effects on the nutritional value and the content of endogenous toxic and antinutritive compounds have to be considered. The potential transfer of allergenic proteins requires special attention. The advisory body is in favor of an authorization procedure comparable to that employed for the safety evaluation of food additives; decisions should be made on the basis of a case-by-case evaluation. Four decision tree structures have been set up in the report for the safety evaluation of simple substances and chemically modified mixtures, foods of vegetable origin, foods of animal origin and foods of microbial origin. It has been proposed to evaluate the usefulness of these decision tree structures after a period of three years.

In the United Kingdom an Advisory Committee on Irradiated and Novel Foods has been appointed in 1984 and reconstituted as Advisory Committee on Novel Foods and Processes (ACNFP) in 1988 to advise ministers and to give guidance to the food industry on the requirements for safety assessment of novel foods and processes. Guidelines on the Assessment of Novel Foods and Processes have been issued by the ACNFP in 1991 (16); a revised and up-dated version of these guidelines has been sent out to organizations for comments in 1994 (17). This committee has also published a report on the use of antibiotic resistance markers in genetically modified food organisms in 1994 (18).

The ACNFP report (16) provides guidance for those producing novel foods or those intending to market them by means of a decision tree scheme, which, by answering a series of questions, indicates the type of information likely to be required in individual situations. Depending on the category, which a novel food or a novel food ingredient belongs to, the systematic questioning approach leads to different exit points. These exit points require certain combinations of information out of the following areas:

I.   Instructions for use
II.   Evidence of previous human exposure
III.   Intake/extent of use
IV.   Technical details of processing and product specification
V.   Nutritional studies

VI.   History of organism
VII.   Characterization of derived strain
VIII. Toxicological assessment
IX.   Human studies
X.   Assessment of genetic modification procedure
XI.   Effect of a genetic modification procedure on the known properties of the parent organism
XII.   Genetic stability of modified organism
XIII. Site of expression of any novel genetic material
XIV.   Transfer of the novel genetic material
XV.   Assessment of a modified organism for survivability, colonization and replication/amplification in the human gut.

In the up-dated version (17) the ACNFP has revised its existing decision tree and has added fifteen new structured schemes, each of which further refines/amplifies the information requirements listed above.   It has been emphasized that the decision tree approach is intended to be applied flexibly and should not be regarded as a rigid checklist.

Under the voluntary arrangements operated in the UK the ACNFP has evaluated several novel foods containing or derived from genetically modified organisms according to its guidelines.   Among these are a genetically modified bakers yeast (19) and a beer produced by a genetically modified brewers yeast (20).

The marketing of these genetically modified microorganisms would require a separate submission according to the Directive on the Deliberate Release into the Environment of Genetically Modified Organisms (*90/220/EEC*) (3) as long as the EU Regulation on Novel Foods and Novel Food Ingredients is not in force.

In December 1994 the ACNFP carried out food safety evaluations on processed products from genetically modified food crops. These are: tomato paste from genetically modified tomatoes, oil from genetically modified oilseed rape and processed food products from genetically modified soy beans. None of these products will be available for sale in the UK until the UK Food Advisory Committee has given its advice on labeling (21).

Literature Cited

1.   Official Journal of the European Communities No. L 117/1, May 8, 1990.
2.   Official Journal of the European Communities No. L 374/1, December 31, 1990.
3.   Official Journal of the European Communities No. L 117/15, May 8, 1990.
4.   Official Journal of the European Communities No. L 230/34, August 19, 1991.
5.   Official Journal ofthe European Communities No.L 214/40, August 24, 1993.
6.   Official Journal of the European Communities No. C 190/3, July 29, 1992.
7.   Official Journal of the European Communities No. C 16/10, January 19, 1994.

8.   Strategies for Assessing the Safety of Foods Produced by Biotechnology. Report of a Joint FAO/WHO Consultation, WHO, Geneva, 1991.
9.   Health Aspects of Marker Genes in Genetically Modified Plants. Report of a WHO Workshop, WHO-Food Safety Unit, Geneva, 1993.
10. Safety Evaluation of Foods Derived by Modern Biotechnology: Concepts and Principles. Organization for Economic Cooperation and Development, OECD, Paris, 1993.
11. Application of the Principles of Substantial Equivalence to the Safety Evaluation of Foods or Food Components from Plants Derived by Modern Biotechnology. WHO Workshop, WHO, Geneva, in press.
12. OECD Workshop on Food Safety Evaluation. OECD, Paris, in press.
13. Demand for Increased Knowledge in the Safety Evaluation of Novel Foods. Institute of Food Technology, Hohenheim University, Germany, 1994.
14. Food and New Biotechnology: Novelty, Safety and Control Aspects of Foods Made by New Biotechnology. Nordic Working Group on Food Toxicology and Risk Assessment, Stockholm, 1991.
15. Advisory Report on Biotechnology-Summary. Netherlands Food and Nutrition Council, The Hague, 1993.
16. Guidelines on the Assessment of Novel Foods and Processes. Advisory Committee on Novel Foods and Processes, Report on Health and Social Subjects 38, HMSO, London, 1991.
17. Guidelines on the Assessment of Novel Foods and Processes. Advisory Committee on Novel Foods and Processes, Revisions to Chapter 4, HMSO, London, 1994.
18. Report on the Use of Antibiotic Resistance Markers in Genetically Modified Food Organisms. Advisory Committee on Novel Foods and Processes, London, 1994.
19. Advisory Committee on Novel Foods and Processes (ACNFP) Annual Report 1989.
20. Advisory Committee on Novel Foods and Processes (ACNFP) Annual Report 1993.
21. Advisory Committee on Novel Foods and Processes (ACNFP) Press Release 1/95, January 19, 1995.

RECEIVED June 7, 1995

Chapter 4

Administrative Oversight To Ensure Safety of Biotechnologically Produced Foods in Japan

Ryoji Takahara

Food Sanitation Division, Ministry of Health and Welfare,
1-2-2 Kasumigaseki, Chiyoda-ku, Tokyo 100-45, Japan

In December 1991 the Ministry of Health and Welfare, Japan, announced its Basic Principle Document for Food Biotechnology, Manufacturing and Safety Guidelines for Foods Derived from Recombinant DNA. The Basic Principle Document covers four biotechnologies: recombinant DNA, cell fusion, tissue culture, and bioreaction. At the present, the Ministry is involved in: 1. Reviewing applications for the Ministry's confirmation which would indicate that the safety assessment conforms with the relevant guideline. 2. Establishing guidelines which cover those technologies which are not covered by the two existing guidelines. Possible new elements in the coming guidelines might be allergenicity of newly introduced proteins and the safety of antibiotic resistance marker genes. 3. Other issues being considered are labelling and the application of the concept of "Substantial Equivalence". Future Strategies and Suggestions for the Ministry to consider are: 1. Deregulation, 2. Risk Communication, 3. Promotion of Information Exchange, and 4. Strategy for Introducing Biotechnology into Food Production.

1. Background

Research and development in biotechnology are making rapid progress in the production and processing of food, aiming at higher quality and better productivity of foods and related products. The Ministry of Health & Welfare, Japan, announced its Basic Principle Document for Food Biotechnology, Manufacturing and Safety Assessment Guidelines for Foods Derived from Recombinant DNA Techniques in December 1991. The Basic Principle Document, established to ensure the safety of consumers, considering the rather short history of biotechnology applications in food production, covers four biotechnologies (recombinant DNA, cell fusion, tissue culture and bioreactors).

Two Guidelines, in effect since April 1992, cover non-novel foods which do not contain recombinant DNA itself. They provide Good Manufacturing Practice and items necessary for safety assessment. They also provide that manufacturers/importers may submit an application to obtain the Minister's confirmation that says their practice conforms with the description in the two guidelines. Our Ministry is urging related industries to apply for the Minister's confirmation at the introductory period of biotechnology in food production, in order to ensure the safety of such foods and to foster their public acceptance.

2. What are we doing now?
1) Reviewing applications for the Minister's confirmation.
Once an application is accepted, a final decision, regarding whether the Minister should confirm that the manufacturing practices/safety assessment methodology of producers/importers conform with the two guidelines, is reached through consultation with the Food Sanitation Committee, which is a scientific advisory board to the Minister. Last December, we received official applications from two companies; these cases are Chymosin derived from *Escherichia coli* K-12 and *Kluyveromyces marxianum* var. lactis. Now we are in consultation with the Food Sanitation Committee; therefore, no confirmation was made. In addition, several companies are preliminarily consulting with the Ministry. The products waiting for submission of the final package of documents are mainly enzymes.
2) Establishing guidelines which cover those technology/products which are not covered by the two existing guidelines.
Manufacturing and safety assessment guidelines for foods 1) derived from recombinant DNA techniques and are novel or with recombinants, and 2) derived from cell fusion, tissue culture, and bioreactor techniques, have not yet been established. We are preparing these guidelines in consultation with scientists.
Major differences with existing guidelines would be provisions regarding allergenicity of newly introduced proteins and safety of antibiotic resistance marker genes.
As for allergenicity, it seems appropriate to ensure that newly introduced proteins produced by biotechnology do not significantly increase allergenicity of the traditional plants. This could be achieved through examining molecular weights, amino-acid structure, stability to heat/acidity/digestive enzymes, immunological response to IgE of the newly introduced proteins and allergenicity of the donor organisms.
As for the safety of antibiotic resistance marker genes, it seems appropriate to ensure that consumption of these products would not significantly increase the antibiotic resistance of gut microorganisms, when considering the antibiotic resistance of microorganisms consumed daily with foods. This could be achieved by examining the potential for horizontal gene transfer of the antibiotic resistance trait from ingested genetically modified plants to gut microorganisms and the potential for inactivation of antibiotics orally administered.

3. Summary of interim report 1993
Safety assessment of foods and food additives produced by biotechnology. In 1991 the study team of Japanese scientists that is one of Food Sanitary Research,

founded by the Ministry of Health and Welfare, commenced its research on rDNA techniques, tissue culture techniques, cell fusion techniques, bioreactor techniques, etc. As it was considered necessary for the development of future guidelines, the study included a fact-finding survey related to foods and food additives produced by biotechnology.

Possible commercialization and marketing in the near future was considered, and research so far identified several issues that needed to be elaborated for the purpose of ensuring the safety of foods and food additives produced by biotechnology, as well as for the development of future guidelines. These items are summarized in this interim report of this study and include the following:

1) Application of the concept of novelty to foods and food additives produced by biotechnology.

2) Safety assessment of food crops produced by rDNA techniques.

3) Safety of antibiotic resistance marker genes used in foods produced by recombinant DNA techniques.

4) Food allergies as related to foods and food produced by biotechnology.

Summary of this report is as follows:

1) Application of the concept of novelty to foods and food additives produced by biotechnology.

1. Introduction
    Since April 1992 the Ministry of Health and Welfare has been enforcing its "Guidelines for Production of Foods and Food Additives Produced by Recombinant DNA Techniques" and "Guidelines for Safety Assessment of Foods and Food Additives Produced by Recombinant DNA Techniques." These guidelines are applicable to foods in which the recombinants themselves are not intended to be consumed and the food products are identical to or deemed to be identical to existing products. These are intended for the cases in which microorganisms are modified by rDNA techniques and used for the production of useful substances such as food additives consisting of enzymes.
    In such products, the probability that the recombinants themselves are consumed depends largely on the methods of extraction and purification of the useful products. The judgment in this respect can basically be supported by several methods such as inspection of products. Assuming no consumption of the recombinants can be attributed to the level of purity, safety assessment should focus on the presence or absence of impurities, and the safety of impurities. If the recombinants may be consumed, it is necessary to consider the safety of the recombinants themselves, in addition to the functions added by recombination. These two cases need to be discussed separately.
    The applicability of these guidelines is limited to products that are identical to or deemed to be identical to existing products, because the necessity for safety

assessment based on toxicity tests varies depending on the presence or absence of novelty and, as such, the safety of the two categories of products should be discussed separately. If a product is identical or deemed to be identical to an existing product, it can be assumed that the safety of the existing food or food additive used as the reference has already been established. If such is the case, we only need to conduct a safety assessment related to the problems added by rDNA techniques and the recombinants.

Since 1989 OECD has been studying measures to ensure the safety of foods and food additives produced by biotechnology. In 1993 it published the "Safety Assessment of Foods Produced by New Biotechnology: Concepts and Principles." In this report the concept of substantial equivalence was proposed to be used for the safety assessment of foods and food additives produced by biotechnology, suggesting that conventional foods should be used as the reference.

The demonstration of substantial equivalence requires:

(1) Knowledge of the composition and characteristics of the traditional or parental product or organism;

(2) Knowledge of the characteristics of the new component(s) or trait(s) derived, as appropriate, from information concerning:

The component(s) or trait(s) as expressed in the precursor(s) or parental organism(s); transformation techniques (as related to understanding the characteristics of the product) including the vector(s) and any marker genes used; possible secondary effects of the modification; and the characterization of the component(s) or trait(s) as expressed in the new organism; and

(3) Knowledge of the new product/organism with the new component(s) or trait(s), including the characteristics and composition [i.e., the amount of the component(s) or the range(s) of expression(s) of the new trait(s)] as compared with the conventional counterpart(s) (i.e., the existing food or food component).

The current Japanese approach, in which the scope of safety assessment concerning food produced by rDNA techniques is limited to foods that are "identical to or deemed to be identical to existing foods," is consistent with the conclusions of the OECD study. However, specific criteria for determining what product is identical to or deemed to be identical to existing foods may differ depending on the properties of actual organisms or products to be modified.

As described above, the Ministry of Health and Welfare's "Guidelines for Production of Foods and Food Additives Produced by Recombinant DNA Techniques" and "Guidelines for Safety Assessment of Foods and Food Additives Produced by Recombinant DNA Techniques" are intended for cases in which microorganisms are modified by rDNA techniques and used for the production of useful substances such as food additives consisting of enzymes. As such, the intended product is generally a simple substance. Therefore, the comparison with existing foods and the judgment of substantial equivalence can be made by comparing chemical structures, sequences, functions, etc.

On the other hand, if crops, animals, or microorganisms are modified and the recombinants themselves are consumed by humans, the products can contain a multitude of different substances. It is thus practically impossible to compare chemical structures, sequences, or functions of each substance. Therefore, the judgment of substantial equivalence should be based on the variation in the contents

of major constituents and the increase or decrease of known toxic substances.
In this report, I first will summarize the principle of "foods or food additives identical to or deemed to be identical to existing foods or food additives" stated in the guidelines of the Ministry of Health and Welfare. Then I will discuss what approach we should take when actually judging whether a product is "identical to or deemed to be identical to existing products" as applied to recombinant crops, which are being developed for commercialization.

2. Principle of "Foods or Food Additives Identical to or Deemed to be Identical to Existing Ones" in the present Ministry's Guidelines.
The guidelines are intended for cases in which microorganisms are modified by rDNA techniques and used for the production of useful substances such as food additives consisting of enzymes, as have already been discussed above.
In specific terms, "identical to conventional" means that identity can be proven on a scientific basis, as summarized on page 100 of the "Guidelines for Foods and Food Additives Produced by Recombinant DNA Techniques" (ed. Food Sanitation Division, Environmental Health Bureau, Ministry of Health and Welfare, Chuo Hoki Shuppan, 1992, ISBN4-8058-0960-4). This publication gives the following definition: "In low molecular weight compounds, their chemical structure should be identical to conventional ones and in higher molecular weight compounds, the chemical structure and sequence of components should be identical."
On the other hand, it states that the concept of "those deemed to be identical" is proper only for higher molecular weight compounds. It continues as follows: "For lower molecular weight compounds, it is difficult to consider that those are identical, because the toxicity of such chemicals is known to change even with a slight difference in structure. Thus, even if the new chemical structure is almost identical to a conventional compound, its toxicological identity cannot be estimated.
"In contrast, some higher molecular weight compounds whose components and functions are identical to conventional ones but whose sequences differ slightly, may be deemed to be identical. For instance, proteins, such as enzymes produced by rDNA techniques, that have functions identical to conventional ones, but slight differences in the sequence of amino acids, can be deemed to be identical to the conventional ones."

3. Principle of "Foods or Food Additives Identical to or Deemed to be Identical to Conventional Ones" as Applied to the Modification of Food Crops and Other Products by rDNA Techniques.
Expecting the modification of food crops and other products by rDNA techniques, the "Guidelines for Foods and Food Additives Produced by Recombinant DNA Techniques " (ed. Food Sanitation Division, Environmental Health Bureau, Ministry of Health and Welfare, Chuo Hoki Shuppan, 1992, ISBN4-8058-0960-4) stated as follows: "Foods and food additives which are complex chemical mixtures produced by rDNA techniques, may be deemed to be identical to the conventional ones if individual chemicals are not changed, even if variations in the relative amounts of these chemicals differ slightly from those of

the conventional ones. Moreover, even if similar new trace compounds are contained, a mixture whose components are not changed and whose levels differ slightly from the fluctuation range of its conventional counterpart may be deemed to be identical."

Based on these principles and the discussion by OECD, it is considered appropriate to take the following position concerning the idea of "identical to or deemed to be identical to conventional ones" as applied to the modification of food crops and other products by rDNA techniques.

1) Premises

(1) Novelty provides a rough indication of the scope and extent of the safety assessment required, although novelty itself does not imply any judgment of safety.

(2) The judgment of novelty should not be based on whether the product is produced by biotechnology but should take into consideration the genetic material and its consumption history, constituents, ingestion method, etc.

2) Applicability

(1) Judgment should be made for each food rather than a group of foods.

(2) Consideration should be given to the normal range of fluctuation of major constituents and other ingredients that occurs in conventional foods and food additives.

3) Specific Items for Judgment

(1) Genetic material

• The type and origin of the host organism and the type and origin of the gene donor used for gene transduction.

• The amount, site, and timing of expression.

• Whether transduction can be achieved by conventional breeding methods.

(2) Extensive experience of safe consumption by humans

• Consideration should be given to the presence or absence of consumption history by not only Japanese consumers but all human consumers, including any consumption of a species related to the newly developed species. However, if:

a. consumption is localized to a group or groups known to be different from Japanese in terms of genetic constitution or metabolism of food constituents, or

b. only limited amounts of the food are consumed on limited occasions, then the consumption history should be considered to have limited value and may be difficult to support the safety of the food.

(3) Major food constituents and other ingredients

• Fluctuation of major constituents and variation in known toxins or antinutrients.

(4) Difference in the method of use between conventional species and new species

• Harvest timing (degree of maturity) and storage method.

• Parts consumed (edible parts).

• Amount consumed

• Methods of cooking and processing.

(5) Other

In ambiguous cases, applicants should be allowed to consult with the Ministry of Health and Welfare.

4) Criteria for Judgment

Agricultural crops and other products produced by rDNA techniques can be considered as products identical to or deemed to be identical to conventional foods, provided that there is no difference from a conventional species in the method of use and if all of the following conditions are satisfied:

a. The donor of the genetic material is identified.

b. The genetic material of the edible parts of the recombinants has been derived predominantly from foods and food additives with a known history of safe consumption by humans.

c. The major constituents of the products are identical to or deemed to be identical to conventional foods.

Products satisfying these conditions may be judged as products lacking novelty, while others may be judged as products with novelty.

5) Other

Although novelty is an important factor affecting the judgment on whether a food is a "newly developed food" as defined in Article 4-2 of the Food Sanitation Law, novelty itself does not directly lead to the judgment that the food is a "newly developed food" defined under the Food Sanitation Law.

4. Other Cases in Which Animals, Microorganisms, etc. are Consumed as Food Modified by DNA Techniques.

In cases in which animals or microorganisms are modified by DNA techniques, it is considered possible to define the range of products that are identical to or deemed to be identical to existing ones in a similar manner using the concept of substantial equivalence. Specific issues should be discussed in detail according to the progress of related techniques in Japan.

2) Safety Assessment of Food Crops Produced by Recombinant DNA Techniques

1. Introduction

The Ministry of Health and Welfare defined and promulgated its "Guidelines for Production of Foods and Food Additives Produced by Recombinant DNA Techniques" and "Guidelines for Safety Assessment of Foods and Food Additives Produced by Recombinant DNA Techniques" in December 1991 and has been enforcing them since April 1992. These guidelines are applied to the production of foods or food additives in which the recombinants themselves are not intended to be consumed and the products are identical to or deemed to be identical to existing products. These are mainly intended for the production of enzymes and other materials at factories using microorganisms as hosts. However, rDNA techniques have also been used in recent attempts to modify food crops and introduce desirable characteristics. Some studies are approaching the stage of commercialization. The current guidelines, as mentioned above, do not cover such modified crops.

Whether rDNA techniques are in fact used or not, when an existing food crop is modified, it is necessary to study any changes caused by the introduced characteristics or the introduction processes and to evaluate safety regarding consumption of the crop as a food or food additive by humans. As we have little experience in the use of rDNA techniques for food production, additional safety evaluations must be conducted on the specific features of rDNA techniques as discussed below.

In this paper I will describe items that need to be considered in safety evaluations of modified food crops, focusing on the changes in crops caused by introduced characteristics and introduction processes, as well as the specific features of rDNA techniques.

2. Products Covered by Safety Evaluations

Safety evaluations should cover food crops (cereals, vegetables, fruits, etc.) that have been modified from existing food crops by the use of rDNA techniques and considered to have no novelty. In addition, the evaluation should cover (1) crops in which introduced genes are expressed in parts that are normally not consumed and (2) crops that are normally processed into food products which are free from proteins and other substances (e.g., rape oil).

3. Method of Evaluation

1) Principles

• Safety evaluation should be conducted by the parties who develop or import the covered products on their own, not by the Ministry of Health and Welfare, as in the cases of all other foods.

• The purpose of presenting basic standards for safety evaluations is to provide general scientific guidance concerning safety evaluations conducted by producers and importers. Such guidance is considered necessary because the use of biotechnology in food production has a relatively short history. This guidance should not be considered as fixed rules, but should be used flexibly according to the development and use of related techniques.

• Safety evaluations should be conducted for individual foods, not for groups of foods. However, evaluation of individual foods does not have much significance when it is applied to items that are already understood scientifically and established, concerning a host, a vector, introduced genes, and preparation of recombinants.

• Safety evaluations should be conducted taking into consideration the site, timing, and amount of gene expression, paying attention to the range of normal fluctuation of main constituents that occur in existing foods and food additives, and using evaluation criteria based on the level of safety of foods which are currently accepted on the market.

• Safety evaluations should be based on the intended method of use in terms of the degree of ripening at the time of harvest and consumption, the method of storage, the parts consumed (edible parts), the amount consumed, and the method of cooking and processing.

• With respect to the results of toxicity tests, the tests should be conducted according to GLP at facilities that conform to GLP. It is desirable that the main parts of the test results are published in academic or similar journals.

2) Basis of Evaluation

The purpose of various breeding techniques is to introduce desirable traits into food crops. Desirable traits can be grouped into three categories: The first is related to production and aims at an increase in yield and harvest efficiency by improving resistance to disease, pests, and herbicides, as well as resistance to drought, cold, and other stress from the growing environment. Second, desirable traits such as improving the processibility and storability of foods are introduced to facilitate processing, transportation, storage, and other stages between production and consumption. Third, improvements related to consumption include improved taste, flavor, texture, etc. to meet the preferences of consumers, as well as modified constituents to remove allergenicity and to enhance nutritional value.

The modification of food crops using rDNA techniques differs from other breeding techniques in several respects. Most importantly, the functions of the transduced genes and vectors have been clarified by detailed scientific study and genes can be introduced from species that are not closely related.

However, there still is a possibility, as in other breeding methods, that a modification of food crops using rDNA technique may cause effects that were not intended. Therefore, evaluations of the safety of modified crops with respect to consumption by humans should include scientific insights into the effects discussed below. The sequence of evaluation need not be as shown here, but evaluations should cover these aspects based on the latest scientific knowledge.

A. Effect of Intended Modification by Gene Recombination

• When the structure or functions of introduced proteins are considerably different from those of existing species, and when the composition of carbohydrates or lipids is significantly altered, it is necessary to identify the changes in the composition, structure, and functions of the constituents and to clarify the effect of these changes on safety for human consumption. Attention should be paid to the alteration of the form in which substances occur, changes in physiological activity, effect on absorption and metabolism (bioavailability) caused by the increase or decrease in antinutrients and other substances that may affect the bioavailability of substances.

• Safety of antibiotic-resistance selective markers (discussed later).

• Allergenicity of newly-introduced proteins (discussed later).

• If the product of the introduced gene or its metabolite is an enzyme, there is a possibility that substances that occur in conventional species may be catalyzed by the enzyme and generate toxic substances. (For example, the catalytic product of the opine-synthesizing enzyme gene is mutagenic.) In addition, an increase or decrease in an enzyme protein caused by modification may lead to accumulation of harmful intermediate compounds or reaction products. Another possibility that needs to be considered is the formation of toxic substances through reaction or interaction with substances produced by other cells, infectious microorganisms, or pesticides and plant growth regulators applied during cultivation.

B. Effect of Unintended Changes by Gene Recombination

• Introduction of unintended traits: pleiotropic expression of the introduced gene.

• There is a possibility that the introduced gene is inserted in active sites of the host chromosomes, resulting in inactivation of existing host genes and alteration of the regulation of gene expression.

• Many plants produce toxic substances and antinutrients to defend themselves from pests and diseases. The levels of these substances in food crops have been reduced so that no toxicity appears after proper cooking. Substitution of chromosomes, insertion of a new regulating domain, and other manipulation can cause reactivation of the metabolic pathways that have degenerated in food crops during the evolution any process, and this may cause the generation of toxic substances and antinutrients. Therefore, especially when such substances are known to occur in the conventional species or species related to it, it is necessary to confirm that the levels of these substances are not increased significantly compared with the conventional species.

3) Specific Items of Evaluation
A. Food or Food Additive Deemed to be Substantially Equivalent (Host)
    • Name, origin and taxonomy: varietal identification, known phenotypes and related genotypes, and genetic ancestors (other species or varieties that provided genetic information to the host in the past).
    • History of safe consumption in and outside Japan.
    • Natural distribution.
    • Population consuming the crop, scale of consuming population, and geographic regions.
    • Amount of consumption, edible parts, methods of cooking and processing.
    • Increase or decrease in toxic substances after cooking or processing.
    • Information concerning growth, harvesting, and storage.
    • Generation of toxic substances, antinutrients, etc. If toxic substances, antinutrients, etc. are known to occur in the host, its ancestral strains, or related species, it is necessary to identify the form in which these substances occur in the modified crop and to confirm the content of these substances.
    • Allergenicity: frequency (epidemiology), causal substances and mechanism, and expected severity of symptoms.
    • Composition of major nutrients and their normal range of fluctuation.
    • Capability and mode of reproduction and genetic properties.
    • Disease, pests, viruses, and other exogenous factors.

B. Vector
    • Name, origin, and the method of preparation/isolation.
    • Structure (restriction endonuclease cleavage map, presence or absence of harmful base sequences, etc.), identification of the part that is actually inserted in the host gene, and its molecular weight.
    • Functions: drug resistance (drugs to which resistance is manifested, mechanism of resistance, residual levels, and metabolites). Transmittability, Host dependence

C. Introduced Gene and Related Items

a) Gene Donor

• Name, origin, and taxonomy: varietal identification, known phenotypes and related genotypes, and genetic ancestors (other species or varieties that provided genetic information to the host in the past).

• History of safe consumption of the gene donor in and outside Japan. Population consuming the crop, scale of consuming population, and geographic regions.

Amount of consumption, edible parts, methods of cooking and processing.

Increase or decrease in toxic substances after cooking or processing.

Information concerning growth, harvest, and storage.

Generation of toxic substances and antinutrients: If toxic substances are known to occur in the gene donor, its ancestral strains or related species, it is necessary to confirm the content of these substances in the modified crop.

Allergenicity: frequency (epidemiology), causal substances and mechanism, and severity of symptoms.

b) Introduced Gene

• Introduced gene related to intended modification

(i) Classification of introduced gene: Attention should be paid to the history of safe consumption of the introduced gene and its product.

a. If the introduced gene is derived from the food crop itself, its use is generally considered safe (except for cases in which the gene is expressed only in inedible parts or the amount of expression is increased significantly).

b. If the introduced gene is derived from crops, animals, or microorganisms that are normally used as food, its use is generally considered safe. (Except for cases in which the gene is expressed only in inedible parts or the amount of expression is increased significantly, but including cases in which, although the introduced gene is derived from species not eaten by humans, the same gene is found in crops, animals, or microorganisms that are normally used as food.)

c. If the introduced gene is derived from any species not eaten by humans, sufficient safety evaluation must be conducted. (Except for cases in which, although the introduced gene is derived from species not eaten by humans, the same gene is found in crops, animals, or microorganisms that are normally used as food, but including cases in which, although the gene is derived from crops, animals, or microorganisms that are normally used as food, the gene is expressed only in inedible parts or the amount of expression is increased significantly.)

(ii) Classification by the function expressed by the introduced gene

a. Special attention is needed when the function of the introduced gene is unknown.

b. The introduced gene is used to code an enzyme protein.

c. The introduced gene is used to code a protein with no enzymatic activity (e.g., gene of virus coat protein).

d. The introduced gene is not translated into protein (e.g., antisense RNA gene).

e. The introduced gene is not transcribed (e.g., insertion of non-coding sequences to destroy a gene).

(iii) Contents of Evaluation

a. Structure (resistance marker gene, promoter, terminator, harmful base sequence, etc.) and origin, function, molecular weight, purity, measurement method and its sensitivity, stability, metabolic pathways (possibility of reaction with substrates that occur in conventional species), presence or absence of exogenous open reading frame, and the possibility of its transcription and expression.

b. The number of copies of the introduced gene.

c. The site, timing, and amount of expression of the introduced gene.

d. Effect of the introduced gene and its product on human intestinal flora.

e. Animal tests confirming the safety of the introduced gene and its product.

f. Allergenicity of the introduced gene and its product (discussed later).

g. Mutagenicity and teratogenicity of the introduced gene and its product.

h. Selective marker gene (discussed later).

i. Method for preparing the expression vector and the insertion site on the host gene (e.g., nucleus, mitochondria, chloroplasts, etc.).

j. Classification of gene and required items of safety evaluation (draft)

D. Recombinants

• Purpose of the use of recombinant plant and the usage as food.

• Approval, situation of use as food, etc. in other countries.

• Method of preparation/cultivation (method for the transduction of the introduced gene into the host, linker, termination codon, selection method, identification method and its sensitivity, method to establish strains for mass-production (e.g., back cross with conventional species).

• Difference from the host.

(i) Mode of reproduction and genetic properties.

(ii) Information concerning growth, harvesting, and storage.

(iii) Difference from conventional species in expected amount of consumption, edible parts, and method of cooking/processing.

(iv) Alteration of metabolic pathways that may affect safety of human consumption: significant changes in substrates, enzymes, and metabolites.

(v) Changes in toxic substances, contaminants, etc.

• Changes in known toxic substances and antinutrients, including those that occur in related species.

• Changes in heavy metal composition: types and quantities.

• HPLC and other analysis of the composition of the recombinant: If the analysis detected a peak or other indications significantly in excess of the normal range of fluctuation in the host, the substance must be identified and its safety must be confirmed.

• Changes in the residual concentration of toxic substances related to changes in the method of cooking/processing (problems related to eating fresh foods, concentration, and denaturation of nutrients, toxic substances, and other major constituents due to application of heat or other processing).

• Residual concentration and safety of chemicals used for selection and their metabolites.

• Animal tests to confirm the safety of the crop or the food that becomes the final product.

(vi) Changes in the composition of major constituents.

If changes in the composition of major constituents are the result of intended modification, evaluations covering the following items should be conducted on the intended changes in the constituents. If the changes are not intended and evidence of safety concerning the changes in the constituents has not been obtained by evaluating other items, the evaluation should cover appropriate items selected from the following, as necessary.

• Changes in the composition of major proteins

If a significant change is detected by two-dimensional electrophoresis, the structure, amino acid composition, quantity, etc. should be characterized and, if possible, the functions should be evaluated.

Newly introduced proteins (origin, biological functions, consumption history, amount of consumption, and method of cooking/processing; in the case of products with no consumption history, chemical homology with edible proteins, denaturation due to cooking/processing, contents in foods consumed by humans, percentage of newly introduced proteins in total protein consumption by humans, allergenicity (discussed later), and host-specific post-transcriptional modification).

If considered necessary based on an evaluation of the above two items, evaluation of non-protein nitrogen content, such as nucleic acids and aminoglycosides, should be conducted.

If a protein or other substance with unknown functions is introduced, protein quality tests should be conducted using the growth of juvenile animals as the index.

• Changes in the composition of major carbohydrates.

If a significant change, percentage of modified carbohydrates in total carbohydrate consumption by humans, is detected in basic composition after hydrolysis, compared with the host, an evaluation covering the structure, composition (sugars, oligosaccharides, polysaccharides, etc.), quantity, etc. should be conducted as necessary.

• Changes in the composition of major lipids

If a significant change is detected in total lipids in the quantitative comparison with the host, an evaluation covering the structure, composition (fatty acid spectrum: phospholipids, sterols, cyclic fatty acids, and saturated, monounsaturated, and polyunsaturated fatty acids), quantity, etc. should be conducted as necessary.

Presence or absence of known toxic fatty acids or fatty acids that are normally absent.

Percentage of modified lipids in total lipid consumption by humans.

• Changes in vitamins, minerals, and other constituents naturally found in the food.

• Alteration of bioavailability due to cooking or processing that alters the form of the above nutrients and due to an increase or decrease in any constituents that may affect the metabolism and absorption of the above nutrients.

• Changes in other constituents.

(vii) Changes in susceptibility to diseases, pests, and viruses.

(viii) Confirmation of the genetic stability of the recombinant: range of fluctuation of the contents of major nutrients, toxic substances, etc.

E. Other necessary information

• If the modified crop is normally eaten after processing, evaluations should cover information about the method of processing, other materials used in processing, processing equipment, refining methods, impurities derived from production processes, concentration and safety of these impurities, changes in constituents due to processing and refining, etc.

• If the product has resistance to herbicides, insecticides, etc., evaluations should cover the mechanism of resistance, the residual levels of these chemicals and their decomposition products, the effect of these substances to human body, etc.

• Information about the contents and safety of other impurities.

• Other items that are considered necessary for safety evaluation when safety has not been confirmed by the above items:

Subacute toxicity test of the product
Chronic toxicity test of the product
Reproduction toxicity test of the product
Carcinogenicity test of the product
Other test results considered necessary for evaluating the safety of the product.

F. Special Remarks

In principle, the functions of the introduced gene should be considered as follows:

If the product for the introduced gene is an enzyme that catalyzes a specific chemical reaction, the function of the gene is considered to be identified by describing the enzymatic properties of the product. However, even if the gene is derived from a species other than the host plant, the safety of the introduced gene may generally be evaluated based on the safety of the host organism (or an organism normally used as food), if provided that both the structure and functions of the introduced gene are considered substantially equivalent to those of the host plant (or an organism normally used as food).

However, in such cases as the introduction of a gene that causes the plant to bear larger fruit, the description of the visible effect of the gene does not clarify the intracellular function of the gene and one cannot forsee the gene's effect to other properties of the host organisms. In this case, the functions of the gene should be considered unknown, and special care should be taken. Even in this case, because the enlargement of the size of the fruit is a modification of the properties of the host plant, this change should be described as the intended modification of the host plant by gene introduction.

3) Safety of Antibiotic Resistance Marker Genes Used in Foods Produced by Recombinant DNA Techniques

1. General Discussions

• A marker gene is usually used to code an enzyme. Therefore, safety should be evaluated in terms of (1) substrate characteristics, (2) safety of the enzymes used as the marker, and (3) safety of enzymatic decomposition products.

• Substrates do not cause any safety problem in themselves, if they are used only in the selection of modified crop cells. However, if the marker enzyme is used in normal farm operations, such as an enzyme with resistance to herbicides or insecticides, the residual levels and safety of these chemicals to human body and their decomposition products must be confirmed with the same level of strictness as in the safety evaluation of ordinary residual chemicals.

• With respect to the general safety of enzymes, no special problem of safety is anticipated unless the enzyme has a peculiar amino acid structure or the amount of consumption is considerably higher than daily protein intake. The safety evaluation of an enzyme should therefore focus on its functions rather than structure.

• There are two types of markers: selective marker genes making use of resistance to antibiotics, herbicides, heavy metals, etc. and screening marker genes used for cell sorting according to the properties of the expression products. The residual levels of these genes, their products, and related metabolites are important factors when evaluating foods, as they are directly consumed orally. It is necessary to accumulate knowledge on the effectiveness of removal and inactivation using biotechnology (the possibility of Cre/Lox site-specific recombination), cooking/processing, and digestion.

• Resistance marker genes also need the same safety evaluation as other inserted genes with respect to toxicity, allergenicity, pleiotropic expression, and other unintended effects that are common to all inserted genes.

2. Safety of Antibiotic Resistance Marker Genes Used in Foods Produced by Recombinant DNA Techniques

1) Principle

• It must be ensured that the ingestion of the recombinant crop does not cause a significant decrease in the effectiveness of orally administered antibiotics associated with the resistance marker.

2) Special Remarks

• When conducting safety evaluations, attention must be paid to the effect in elderly persons, infants, patients with digestive tract disorders, and other cases in which the functions of digestion and absorption have been compromised.

• When conducting safety evaluations of an antibiotic resistance marker gene and its products contained in a single crop, the possibility that cumulative ingestion may increase in the future as more crops are developed with these techniques must be considered.

3) Safety problems related to ingestion specific to antibiotic resistance marker genes

A. Effect of genes and their products on human intestinal flora (horizontal gene transfer from plants to microorganisms)

• While selective marker genes are incorporated in plant genes, the demonstration of gene transfer from plants to microorganisms has not been reported in the literature.

• Gene fragments generated in the process of digestion are readily decomposed in the intestines in a very short period of time, and incorporation into microorganisms or human intestinal epithelial cells does not occur unless the fragments have homology over at least 20 bases. Such incorporation is thus unlikely to occur.

B. Possibility of inactivation of antibiotics (inhibition of therapeutic effect) due to gene products remaining in the intestines and genes transmitted to and expressed in human intestinal epithelial cells.

• Gene products are decomposed rapidly in human intestines, and the environment in the human digestive tract often lacks ATP and other factors needed for the manifestation of resistance.

• There is little possibility that resistance genes will be transferred to human cells. In addition, promoters and other mechanisms involved in gene expression in humans are different from those in plants and microorganisms. Therefore, even if transfer occurs, the possibility that resistance will be manifested is extremely low.

C. Possibility of inactivation of antibiotics added to animal feed

• The problems pertaining to the safety of meat, milk, and milk products are not considered in this study.

4) Items requiring evaluation

(i) Characteristics of the host (amount of consumption, method of cooking/processing, composition).

(ii) Characteristics of genes and gene products.

• Structure and functions of genes and gene products, especially the homology with human DNA, etc.

• Mechanism of the manifestation of resistance, the method of use, and related metabolites: in particular, whether ATP and other co-factors are needed for the manifestation of resistance and the concentration of these substances in the digestive tract.

• Method for the identification and quantitation of genes and their products.

• Changes in genes and their products caused by cooking/processing (stability against heat and physical pressure).

• Changes in genes and their products in the digestive tract environment (stability against acids and digestive enzymes).

(iii) Safety evaluation

• Expected amount of consumption of genes and their products.

• Present usage of antibiotics associated with resistance (quantity and method of use).

• The amount and characteristics of antibiotic resistant bacteria resulting from the consumption of crops with rDNA, as compared with the amount and characteristics of antibiotic resistant bacteria that are normally present and ingested.

• Estimated amount of antibiotics inactivated after oral administration and the possibility that inactivation may cause problems.

4) Food Allergies as Related to Foods and Food Additives Produced by Biotechnology

1. Principles

• If it can be confirmed that protein introduced by biotechnology does not act as an allergen and the gene modification does not cause a significant increase in known allergens in the host, the safety concerning the allergic reaction of the recombinants is considered to need no special attention.

• If protein introduced by biotechnology significantly enhances the antigenicity in the host, adequate measures should be taken to prevent unintentional consumption by sensitive population, paying attention to the degree of antigenicity, the expected severity of allergy, and the scale and characteristics of the population susceptible to allergy (attention should be paid to the effect in elderly persons, infants, patients with digestive tract disorders, and other cases in which the functions of digestion and absorption have been compromised).

• Consumers purchase, cook, and eat their food. Therefore, public relations activities are extremely important. Consumers should be fully informed of the possible consequences of ingesting food containing substances that can cause allergies so with accurate knowledge they can make correct judgments.

2. Relationship between proteins in recombinants and allergenicity
   1) Introduction of known allergens
   This includes cases in which allergenicity is increased as the result of the increase of a particular substance, such as a protein, related to gene recombination in crops.
   2) Items requiring examination with respect to the possibility that newly introduced protein may have allergenicity.
   A. Properties of newly introduced protein
   • Homology search for structural and other similarities to known allergens: molecular weight and amino acid sequence, protein structure (allergenic determinants/IgE binding sites).
   • Xenogeneic characteristics of newly introduced protein in humans.
   • Physicochemical properties of the newly introduced protein, i.e., effect of cooking/processing and digestion (heat stability and acid stability) and reactivity to proteases should be tested for sensitivity to heat treatment, artificial gastric juice, and artificial intestinal juice using protein electrophoresis and the Western blot method. The judgment should be based on:
   (i) whether electrophoresis shows the fragmentation of the introduced protein to low-molecular-weight components (MW < approx. 1,000), and
   (ii) whether positive bands are detected in the Western blot method using animal (e.g., rabbit) antibodies against the introduced protein (or a significant decrease in the binding capacity of antibodies is identified.)
   The samples used in the analysis need not be purified. However, it must be ensured that untreated newly introduced protein can be detected by protein electrophoresis and the Western blot method.
   B. Amount and form of ingested newly introduced protein
   • Content in food (whether the amount ingested of this protein changes significantly).
   • Amount ingested of the host (whether it occupies a significant part in the daily protein intake), its edible parts, timing of harvesting, method of storage, and method of cooking/processing.
   • Consumption history of the gene donor, amount ingested, edible parts, timing of harvesting, method of storage, and method of cooking/processing.
   C. Binding capacity to IgE antibodies against representative allergens and existing allergens with confirmed structural homology (based on the ELISA and Western blot methods).

The search for the binding capacity to IgE antibodies using ELISA method should be conducted paying attention to the following points.
(i) The samples used in the analysis need not be purified completely. However, because of the characteristics of the ELISA method, the degree of purity of the newly introduced protein in the sample may have an influence on the measurement results (coating of the plate with the sample). Therefore, newly introduced protein must be purified when necessary, and the measurement system must be validated using positive and negative controls.
(ii) Sera from patients with food allergy who are sensitive to representative allergens should be used as the antibodies. The number of the testing materials should be from 10 to 20 for each allergen.

4. Other Issues to be Considered
    1) Labeling
    We believe that a general labeling of genetically modified foods cannot be justified by the Food Sanitation Act. If labeling is requested, it should be based on public health rationale, such as a warning that known allergens are contained.
    2) Novelty
    Novelty of foods, which determines the type of guidelines to be applied, should be judged based on "Substantial Equivalence" principle, considering genetic materials used and composition of the product. It seems appropriate that the Ministry announces the basic principle to judge "novelty" of foods, including consulting with researchers.
    3) Range of documents necessary for application requesting Ministry's confirmation.
    We are planning to clarify the range of documents necessary for application requesting Minister's confirmation, in addition to strengthening of the administrative capabilities for a prompt consultation.

4) Others
    When importers submit an application requesting the Minister's confirmation which states that their methodology of safety assessment of foods derived from biotechnology conforms with the relevant guideline, it takes time to translate scientific materials into Japanese language and to inquire into issues to be clarified to their head offices abroad. It would expedite procedures to provide Ministry with advance notice regarding the range of documents which should be translated into Japanese.

5. Future Strategies and Suggestions
    1) Deregulation
    As safety data on foods derived from biotechnology accumulates, it seems appropriate to ease, as necessary, manufacturing/safety assessment guidelines and limit the range of scientific documents to be submitted in order to obtain the Minister's confirmation of conformity to the two guidelines.
    2) Risk Communication
    In order to increase public understanding/acceptance and consumer

confidence in foods derived from biotechnology, it seems vital to involve the public in scientific discussions on the safety of such foods and to provide them with appropriate information in an understandable manner.

3) Promotion of Information Exchange

In order to best utilize limited resources in an efficient and timely fashion, it is important to share information on the safety of foods derived from biotechnology amongst industries, academia, regulatory authorities, and the public in Japan and abroad.

4) Strategy for Introducing Biotechnology into Food Production

In order to facilitate the public acceptance of foods derived from biotechnology, it seems useful to consider a strategy for introduction of such foods into the market, namely, which product comes first, based not only upon technological feasibility, but also on placing priority on foods which are easy to accept and for which benefits are visible for consumers. Such foods would include, for instance, a plant for which allergenicity is reduced by biotechnology, which does not contain antibiotic resistance marker genes, or which contains higher concentration of iodine, etc.

6. Additional Thoughts

When we consider

1) the expected gap between the population increase world-wide and the limitation of area for cultivation and

2) the necessity of environmental protection by reducing the use of pesticide chemicals which could be achieved through introducing disease resistant crops derived from biotechnology, it becomes clear that applications of biotechnology in food production bring about benefits, not limited to the increase in productivity/ quality of foods.

In contrast to the use of pharmaceuticals in which side-effects are tolerated, biotechnology in food production suffers from the lack of tangible benefits to consumers and from the higher level of safety assurance procedures required. It also seems improbable that the price of newly developed foods falls significantly short of that of their traditional counterparts.

When this gap between the necessity and the hindrance of application of biotechnology into food production is carefully examined, one could conclude that the importance of safety assurance regarding such foods, emphasizing long-term and wide-ranging benefits of such foods, and communication of their utility to the public at every opportunity, including a symposium like this, can never be overemphasized.

RECEIVED July 14, 1995

Chapter 5

The Role of the Public and Federal Advisory Committees in Providing Advice to the Government on Agricultural Science Policy

Alvin L. Young and Daniel D. Jones*

Office of Agricultural Biotechnology, U.S. Department of Agriculture, Room 1001, Rosslyn Plaza East, Washington, DC 20250–2200

Advisory committees have played an important role in shaping programs and policies of the Federal government from the earliest days of the United States. Today, an average of 1,000 advisory committees with more than 20,000 members advise the President and the Executive Branch on such issues as genetic engineering, agricultural research priorities, standards for organically produced foods, and the safety of food ingredients. With the expertise from advisory committee members, Federal officials and the Nation have access to information and advice on a broad range of issues affecting Federal policies and programs. The public, in turn, is afforded an opportunity to participate actively in the Federal government's decisionmaking process.

The development of agricultural biotechnology comes at a critical juncture for United States agriculture. Competition for world markets is growing more and more intense. At the same time, U.S. producers and processors must meet the challenge of new environmental policies. Continued Federal investment in agricultural research is mandatory if the U.S. is to maintain its agricultural standing. Many recent breakthroughs have resulted from the Federal investment in research on genetics, molecular biology, and cell culture.

Indeed, the genetic modification of plants, animals, and microbes, often through advances in biotechnology, is becoming increasingly important in agricultural research and production. This has raised issues of health, safety, and the environment that face both the scientific community and public policy makers. For example, a 1987 National Academy of Sciences report concluded that "the scientific community urgently needs to provide guidance to both investigators and

---

*The authors are the Director and Deputy Director, respectively, of the USDA Office of Agricultural Biotechnology. The views presented are those of the authors and do not necessarily represent official policy or interpretations of the USDA.

regulators in evaluating planned introductions of modified organisms from an ecological perspective" (1).

One mechanism that has proven to be useful both in organizing input from the scientific community and in building public confidence in scientific research is the Federal advisory committee mechanism. With the expertise from advisory committee members, Federal officials and the Nation have access to information and advice on a range of scientific and other issues affecting Federal policies and programs. The public, in turn, is afforded an opportunity to participate actively in the Federal government's decisionmaking process.

The United States Department of Agriculture (USDA) plays a number of roles in agricultural biotechnology, supporting and encouraging various initiatives and ensuring that biotechnology is developed safely. Ten USDA agencies participate in biotechnology activities -- the Agricultural Research Service, Cooperative State Research Education and Extension Service, Extension Service, Forest Service, Animal and Plant Health Inspection Service, Food Safety and Inspection Service, Agricultural Marketing Service, Economic Research Service, National Agricultural Library, and The Foreign Agricultural Service.

Scientific Advice for Agriculture

USDA strongly supports basic and applied research in biotechnology, allocating $180 million for the effort in Fiscal Year 1994. Some $30 million additional funds were allocated in providing facilities and equipment at universities and Federal laboratories and in the maintenance of repositories and data bases. Decisions on research priorities and special projects and initiatives require the best advice from the agricultural and environmental communities and from the general public.

In order to secure the very best scientific advice for the Department on agricultural biotechnology research and related issues, USDA established a scientific advisory committee in 1987 called the Agricultural Biotechnology Research Advisory Committee (ABRAC). The ABRAC was modeled in part after the National Institutes of Health Recombinant DNA Advisory Committee (NIH-RAC) which reviewed recombinant DNA research primarily in medicine and the health sciences. The ABRAC differs from the NIH-RAC in that it embodies expertise and experience specific to the agricultural research community.

Purpose of the ABRAC

The purpose of the ABRAC is to provide advice to the Secretary of Agriculture on policies, programs, operations, and activities associated with the conduct of agricultural biotechnology research. This includes administrative and procedural measures designed to promote the safety, effectiveness, and public acceptance of agricultural biotechnology research.

The ABRAC oversees the review of proposed research projects, evaluates the adequacy of draft proposals used by USDA in preparing environmental assessments of research projects, recommends necessary revisions to research guidelines and protocols, and advises other Federal and State agencies on agriculturally related research projects. It can also provide information to and maintain cognizance of institutional biosafety committees (IBC's) to assure the availability of essential personnel to carry out oversight of agriculturally related biotechnology functions. The ABRAC, when it is asked to do so, can also help to strengthen the scientific basis and credibility of actions taken by regulatory agencies within USDA.

The ABRAC is cited for consultation by the Secretary in the 1990 Farm Bill, Section 1668, on Biotechnology Risk Assessment Research. As a result of this legislation, the USDA administers a small biotechnology risk assessment research program funded at a level of one percent of the Department's total expenditures on biotechnology research. In Fiscal Year 1994, the amount of funding for the biotechnology risk assessment research program was approximately $1.7 million.

Composition of the ABRAC

The ABRAC consists of 15 members drawn from academia, industry, and government with knowledge in such fields as animal/veterinary science, fisheries science, plant science, forestry, microbiology, food science, environmental science and policy, laws and regulations, and bioethics.

The rotating membership of the ABRAC provides an opportunity for changing the mix of experts on the Committee to accommodate the evolution of issues facing USDA. For example, the ABRAC now includes experts in fisheries science and food science that were lacking in the first ABRAC. The appointment of ABRAC working groups also provides the flexibility to recruit experts outside the ABRAC to address biotechnology issues as they emerge and evolve. For example, an ABRAC Transgenic Animal Working Group included molecular biologists and animal scientists from outside the ABRAC. Because the ABRAC is a Federal advisory committee, transcripts and minutes of meetings are prepared. The minutes are published and their availability to the public is announced.

Accomplishments of the ABRAC

ABRAC accomplishments include the review of individual research proposals involving questions of biosafety, the development of guidelines and performance standards for research with genetically modified organisms, and, in response to a specific request, the development of recommendations for the food safety evaluation of transgenic livestock.

The ABRAC reviewed several individual research proposals for biosafety including proposals involving transgenic carp, transgenic catfish, and genetically modified strains of *Rhizobium* and *Brucella*. Several of these proposals, because of potential impacts on the environment, involved the preparation of environmental assessments under the National Environmental Policy Act. For example, the ABRAC advised the former USDA Cooperative State Research Service (CSRS), which funded extramural research, on two controversial proposals for studies of transgenic fish in outdoor research ponds. Based on ABRAC recommendations and appropriate environmental assessments, CSRS approved studies for transgenic carp in 1990 and transgenic catfish in 1992.

In 1991-92, the ABRAC recommended to USDA a set of practices and procedures for the safe conduct of research involving the planned introduction into the environment of certain genetically modified organisms (2). The guidelines established principles for assessing the safety of research with specific organisms and designing confinement measures to promote safety. The guidelines were intended to aid researchers and institutions in the design of safe experiments conducted outside contained facilities. The ABRAC also provided eight specific examples to illustrate how the guidelines would be applied to different modified organisms; these were cattle, carp, pine, rapeseed, fruit fly, a parasitic wasp, and two species of bacteria, *Pseudomonas fluorescens*, and *Clavibacter xyli*. The ABRAC is currently working with the aquatic research community and state and Federal agencies on the development of standards of confinement for outdoor research on genetically modified fish and shellfish.

The ABRAC, when requested, has also assisted regulatory agencies by providing scientific advice on specific issues or proposals facing them. For example, the Food Safety and Inspection Service consulted the ABRAC on its proposed policies for both transgenic animals and non-transgenic animals from transgenic experiments. These consultations resulted in *Federal Register* notices concerning transgenic animals (3) and non-transgenic animals from transgenic experiments (4). The USDA Animal and Plant Health Inspection Service has also worked with the ABRAC on the scientific basis for performance standards for field testing of selected crops.

Public Acceptance of Scientific Research

Public acceptance of biotechnology research as well as the products of biotechnology is emerging as a critical issue for the 1990's. One of the greatest strengths of the ABRAC is the broadly based and open process that it provides for surfacing special interest concerns about biotechnology and examining them in a public forum and in the light of scientific knowledge. Thus, ABRAC is the primary mechanism the Department uses to promote dialogue with the public and forge a public policy consensus on controversial issues pertaining to the development of agricultural biotechnology.

One area which exemplifies the ABRAC's role in promoting public acceptance of biotechnology research is research on genetically modified fish and shellfish. Environmental and other interest groups have repeatedly called for improved procedures and standards for research on genetically modified organisms in aquaculture research. This could also reduce the need for resource-intensive, case-by-case evaluations of the safety of research proposals in the future.

The ABRAC has been involved in a broad-based effort including the aquatic research community, the aquaculture industry, and Federal and state government agencies to develop generic standards for assuring the safety and environmental friendliness of aquatic research with genetically modified fish and shellfish. This broad-based approach was adopted to include as many stakeholders in the process as possible and to increase the likelihood of general acceptance of the standards.

The ABRAC therefore formed a Working Group on Aquatic Biotechnology and Environmental Safety which included aquaculture and aquatic research experts from outside the ABRAC. The Working Group met in conjunction with a workshop for scientists, business people, government officials, and other interest groups at the University of Minnesota in 1993. Workshop participants drafted performance standards to help scientists with the ecological risk assessment of research involving genetically modified fish, crustaceans, and molluscs. Final ABRAC recommendations on scientific protocols to help assure the environmental safety for outdoor research on modified fish and shellfish are expected in the near future.

ABRAC also plays a significant role in promoting international dialogue on biosafety issues. The Committee provided an external review mechanism for the Organization for Economic Cooperation and Development's document on the safety of large-scale testing of crop plants.

Biosafety Results of Field Tests

A meeting of the ABRAC was held in tandem with the 3rd International Symposium on the Biosafety Results of Field Tests of Genetically Modified Plants and Microorganisms in late November, 1994 in Monterey, California. Questions about the ecological effects of field testing were discussed, especially in those countries where large-acreage field tests are in progress with less oversight than in the U.S. The ABRAC has environmental science and policy expertise to address most questions relating to environmental impacts of agricultural field tests. Some of the relevant questions discussed at the November meeting included:

o   Are field testing risks scale-dependent?

o   Can small-scale results be extrapolated to larger scale?

o   What are the long-term effects, if any, of agricultural field testing?

o   Are there unique risks when testing genetically modified organisms in natural centers of diversity?

o   Are there unresolved issues regarding the possible generation of new viral pathogens from transgenic plants?

Organic Foods and Biotechnology

Another issue which may become important for USDA in the future is the possible use of biotechnology in the production of organic foods. USDA, under the 1990 Farm Bill, administers a National Organic Standards Board which is addressing a broad range of questions concerning definitions and standards for organic foods. One of these questions is whether recombinant DNA techniques can be used in the production of foods which are represented to the public as organically produced. The ABRAC, if requested, could serve as a supplementary source of expertise in biotechnology to address scientific aspects of organic standard questions and to develop recommendations for the USDA if needed.

Food Safety and Public Acceptance

The safety of emerging food products of biotechnology is a critical concern of special interest groups and the public. Recombinant chymosin, bovine somatotropin, and the FLAVR SAVR tomato are the first of many food products of biotechnology, some controversial, to enter the food supply. USDA has a strong interest in encouraging public acceptance of any new food products that are safe, wholesome, and properly labeled. ABRAC is beginning to address the acceptance to the public of the use of genetically modified organisms in agricultural research and production. The ABRAC provides a useful public forum in which concerns about the definition and safety of new food products, especially those regulated by USDA, can be raised and discussed in a scientific setting. Issues include:

o   Do transgenic animals expressing human proteins for pharmaceutical use pose unique risks or other problems when presented for entry into the human food chain?

o   Does classical toxicology offer a useful perspective in assessing the safety of whole food products produced by biotechnology?

Societal/Bioethical Aspects of Biotechnology

In 1993, the ABRAC held a meeting in tandem with a conference on the societal issues of food biotechnology (5). While this conference was timely and informative, food biotechnology issues have not gone away and they may arise again in connection with societal impacts of specific new products of biotechnology, acceptance by the public and food service professionals, and impacts on rural economies. The ABRAC can bring logical, philosophical, and bioethical expertise to bear on many of these issues and make a strong contribution to the societal acceptance of new products and technologies.

Future Issues

In addition to the issues described above, other issues in agricultural biotechnology have been considered for possible ABRAC attention. Specific issues in plant science included management of resistance to biopesticides, spontaneous assembly of infectious agents from molecular subunits, production of pharmaceuticals in plants and animals, and the use and effects of synthetic sequences in plants, animals, and microorganisms.

Broader issues included risk management and risk communication in agricultural biotechnology, public attitudes and perceptions, support for biomonitoring databases, and trade constraints to the movement of genetically engineered crops and foods. How many of these issues are placed on the ABRAC agenda will depend in part on resources available to the Committee and on the changing mix of biotechnology issues facing the Department of Agriculture.

Literature Cited

1.   National Academy of Sciences Committee on the Introduction of Genetically Engineered Organisms into the Environment, *Introduction of Recombinant DNA-Engineered Organisms into the Environment: Key Issues*, National Academy Press, Washington, D.C., **1987** .

2.   Agricultural Biotechnology Research Advisory Committee, Supplement to Minutes: *Guidelines for Research Involving Planned Introduction into the Environment of Genetically Modified Organisms*, Document No. 91-04, U.S. Department of Agriculture, Washington, D.C., **1992**.

3.   Food Safety and Inspection Service, "Update on Livestock and Poultry Connected With Biotechnology Research," *Federal Register* , 59, (52), p. 12582, March 17, **1994**.

4.   Food Safety and Inspection Service, "Livestock and Poultry Connected with Biotechnology Research," *Federal Register*, 56, (249), p. 67054, December 27, **1991**.

5.   *Symbol, Substance, and Science: The Societal Issues of Food Biotechnology*, Conference Proceedings, June 28-29, **1993**, Burke, W.S., Ed., North Carolina Biotechnology Center, Research Triangle Park, North Carolina.

RECEIVED June 7, 1995

Chapter 6

Safety Assessment of Flavor Ingredients Produced by Genetically Modified Organisms

J. B. Hallagan[1] and R. L. Hall[2]

[1]Law Offices of Daniel R. Thompson and [2]Flavor and Extract Manufacturers Association, 1620 I Street, N.W., Suite 925, Washington, DC 20006–4005

Flavor is a primary determinant for food choice. Food flavors are complex mixtures of individual flavor ingredients which are often natural constituents of food, and can be produced through physical means, chemical synthesis, or more recently, through modern biotechnology. Biotechnology processes may include fermentation, enzymolysis, and cell and tissue culture, with and without the use of genetic modification. The use of genetically modified organisms (GMOs) presents certain issues relevant to the safety assessment of flavor ingredients. The longstanding safety assessment program of the FEMA Expert Panel has been revised to account for the production of flavor ingredients through the use of GMOs so that such flavor ingredients can be evaluated for status as generally recognized as safe (GRAS).

Flavor is a primary determinant for food choice (1,2,3). Changes in the composition of many commonly consumed foods and beverages to meet dietary needs (e.g., low fat, low calorie, and low sodium) have led to significant new demands on flavor producers (3). In meeting these demands, the flavor industry has grown such that annual industry sales were estimated at $5 billion worldwide (4). In recent years, the flavor industry has begun to explore the use of modern biotechnology (e.g., tissue culture and recombinant DNA techniques) to meet the needs of its customers. This report will address the safety assessment of individual flavor ingredients that are produced through the use of genetically modified organisms (GMOs).

Food Flavors

Flavor was defined as "the sum of those characteristics of any material taken in the mouth, perceived principally by the senses of taste and smell, and also by the

0097–6156/95/0605–0059$12.00/0

general pain and tactile receptors in the mouth, as reviewed and interpreted by the brain" (5). The material that imparts the perceived flavor is often a complex mixture of individual substances even though a single substance within the mixture may be the predominant source of the flavor. For many years, flavors added to foods were often extracts and essential oils derived from spices, fruits, and vegetables. The modern flavor industry has developed a catalog of the constituents of natural extracts and oils, and their synthetic counterparts (6); these are the individual flavor ingredients which serve as the building blocks of flavors.

Flavors specifically designed to meet the varying needs of the consuming public can be created by the flavorist using the myriad individual naturally-derived or synthetic flavor ingredients. There are over 1700 materials listed as generally recognized as safe (GRAS) for use as flavor ingredients in the United States by the Flavor and Extract Manufacturers Association (FEMA). The substances listed by FEMA include many individual flavor ingredients that can be produced synthetically, or derived from natural products by various means including fermentation, enzymolysis, heating, and fractionation; some complex mixtures such as natural extracts and essential oils are also considered "FEMA GRAS" because of their long history of safe use in foods. Fermentation systems employing unmodified organisms have long been recognized as sources of flavor ingredients such as fatty acids, methyl ketones, carbonyl compounds, lactones, and esters (7). A variety of bacteria, yeast, and fungi have been identified as useful organisms for the production of flavor ingredients (8,9,10) and genetic modification may facilitate the development of new systems useful in industrial production. An extensive group of flavor ingredients have been produced by various microorganisms (11).

The Production of Flavors Using Modern Biotechnology

Flavors are often complex mixtures of individual ingredients, some of which impart flavor and others that serve specific non-flavoring functions such as carriers, preservatives, and emulsifiers. The flavor industry has focused on using modern biotechnology to produce flavor ingredients that impart flavor leaving the production of most non-flavor ingredients to conventional methods. The flavor industry has explored two areas of biotechnology for the production of flavor ingredients: (1) plant tissue culture (with or without genetic modification), and (2) GMOs (i.e. microorganisms) (11,12,13,14,15). A third area is an indirect application of modern biotechnology to the production of flavor ingredients, the use of enzymes produced through genetic modification in conventional food processing such as cheese production, and the subsequent use of various cheese-related products to produce flavor ingredients. Safety issues associated with such enzymes have been thoroughly addressed by FDA in its regulation of enzymes such as chymosin (16). At the present time, the use of GMOs in combination with fermentation and enzymolysis hold the most promise for the production of flavor ingredients through either direct expression or the bioconversion of appropriate substrates; this paper will be limited to comments on the production of flavor ingredients through these techniques even though many aspects of the safety assessment of substances produced by GMOs apply equally well to substances produced by plant tissue culture. Plant tissue culture has not yet become a significant method of production of flavor ingredients because of limited capabilities, difficult purification, low yields, and therefore limited financial rewards. For example, extensive research has been accomplished on producing vanilla products through tissue culture techniques but at this time, tissue culture cannot compete financially with a large and relatively inexpensive supply of vanilla beans. Furthermore, the GRAS assessment program described here for products of biotechnology is limited to the evaluation of individual flavor ingredients and tissue culture techniques are more likely to produce complex mixtures that more closely resemble a plant extract or essential oil. The GRAS assessment of complex mixtures presents other issues which will be addressed in a later report.

FDA's Regulation of Recombinant Chymosin

Substances useful in food processing can be produced by certain microorganisms by fermentation starting with simple nutrients such as sugars or amino acids. For example, the genetic modification of bacteria, fungi, and yeast such as *E. coli, A. niger, and K. marxianus* has been accomplished so that they produce the enzyme chymosin which has significant uses in food processing and has been determined to be GRAS by FDA (17,18,19). Industrial enzymes, including chymosin, produced through the use of GMOs have achieved significant market share possibly in excess of 50% (20).

In its safety review of recombinant chymosin, FDA identified three issues for the evaluation of products produced by expression from genetically modified microorganisms: (1) that the correct gene was cloned and expressed, (2) that the product is substantially similar (chemically and biologically indistinguishable) from the conventionally produced substance, and (3) that any impurities do not result in safety concerns. Following FDA's action on chymosin, an FDA representative commented that the agency's review "of the safety, purity, and identity of these products is fundamentally no different from that of analogous products derived from unmodified or traditionally modified organisms." (16). The three main concerns identified by FDA in its review of recombinant chymosin serve as a basis for the safety review of flavor ingredients and other food constituents produced through the use of GMOs.

The FEMA GRAS Assessment Program for Flavor Ingredients

Private GRAS assessments under Section 201(s) of the Federal Food, Drug, and Cosmetic Act is an approach viewed favorably by both industry and regulators for food ingredients, including flavor ingredients. In its 1992 policy statement on food biotechnology, FDA acknowledged that the private determination of GRAS status for food ingredients is a valid approach.

> ...Congress recognized that many substances intentionally added to food do not require a formal premarket review by FDA to assure their safety...FDA has traditionally encouraged producers of new food ingredients to consult with FDA when there is a question about an ingredient's regulatory status, and firms routinely do so, even

though such consultation is not legally required. If the producer begins to market the ingredient based on the producer's independent determination that the substance is GRAS and FDA subsequently concludes the substance is not GRAS, the agency can and will take enforcement action to stop distribution of the ingredient and foods containing it on the ground that such foods are or contain an unlawful food additive. (21).

The FEMA GRAS assessment program was established in 1960 with the creation of the FEMA Expert Panel, an independent group of experts that, while sponsored by FEMA, operates independently from the association (22). The Expert Panel applies a thorough, critical, and balanced approach to safety assessment and has been recognized by FDA as producing "reliable industry GRAS lists" (23). The Panel's five main criteria for GRAS assessment are:

(1) Exposure to the substance in specific foods, the total amount in the diet, and the total poundage, (2) natural occurrence in food, (3) chemical identity (including purity and method of preparation) and specific chemical structure, (4) metabolic and pharmacokinetic characteristics, and (5) animal toxicity (24,25).

A critical part of the Panel's safety assessment, and therefore its assessment of GRAS status, is the use of analysis by chemical structure analogy; an individual flavor ingredient is analyzed within its class of structurally related compounds. Relevant toxicology, metabolism, and exposure data from structurally related substances are employed to aid in the GRAS assessment of individual flavor ingredients. The Panel will also use relevant data from structurally related compounds even if they are not flavor ingredients. Finally, the Panel's conclusion that a flavor ingredient is GRAS is predicated on the substance's conditions of intended use as a flavor ingredient to be added to foods or beverages at certain specified levels. If the eventual use of the flavor ingredient falls significantly outside of the Panel's specified conditions then the use of the substance is no longer GRAS and therefore, any food or beverage it is added to is adulterated under the Federal Food, Drug, and Cosmetic Act.

Recent developments in the production of flavor ingredients using techniques employing GMOs raise issues relevant to the safety assessment of such flavor ingredients, and the FEMA Expert Panel has modified its GRAS assessment program to account for these new production methods. Issues associated with the safety assessment of food and flavor ingredients derived from processes using GMOs were addressed in a critical and comprehensive manner by the International Food Biotechnology Council (26). The IFBC findings and recommendations, together with FDA policy statements and procedures, were used by the FEMA Expert Panel in the formulation of its policy for the GRAS assessment of flavor ingredients derived from GMOs. The Panel's assessment begins with an evaluation of the identity of the substance at issue including an examination of its regulatory status - is it currently approved for use in food? If so, what are the specifications for the substance and are they met by the new material produced through a GMO?

If the substance produced through a GMO is the same as the conventionally produced substance then the safety assessment is greatly simplified, and in fact is essentially complete if the substance is currently approved for use in food by virtue of the existence of an appropriate prior safety assessment.

The Panel's analysis consists of several levels. The first level consists of a safety evaluation based largely on an examination of the identity of the substance and the method of production employing the genetically modified organism. Within this level, the Panel first evaluates the flavor ingredient itself including current regulatory status, specifications for the substance as produced using a GMO, and any safety considerations associated with constituents of the flavor ingredient as derived from a GMO (Figure 1). As a second level of review, the Panel evaluates specific issues associated with genetic modification such as relevant characteristics of the GMO including whether the microbe ends up in the flavor ingredient, whether there are antibiotic resistance concerns, and whether there are safety concerns associated with the vector and DNA insert code (Figure 2 and Table 1). The Panel's evaluation is performed within the context of its usual inquiry into the method of production of a flavor ingredient. For example, the Panel will examine the method of production of a conventionally produced substance to ensure that no deleterious contaminants are present in the final product. In general, the issues evaluated by the Panel are consistent with the main issues identified by FDA in its evaluation of recombinant chymosin.

Once the Panel has resolved issues associated with the GMO used in the production of the flavor ingredient, it applies the criteria employed in a GRAS assessment as described by Woods and Doull (24) if it is necessary to do so, such as when the flavor ingredient is not currently approved for food use through an appropriate safety assessment. If the Panel determines that the flavor ingredient is GRAS, then the results of its evaluation are published in the typical manner (27). In the Panel's view, the safety assessment of a flavor ingredient produced by modern biotechnology should be no different from the safety assessment of a conventionally produced flavor ingredient once issues associated with the GMO are resolved. The Panel's approach is largely consistent with the "vertical" approach advocated by Miller and Gunary (28).

Conclusion

The production of flavor ingredients using genetically modified organisms presents new safety assessment issues. It is critical to evaluate various aspects of the production process related to the genetically modified organism in order to know what questions to ask before performing a safety assessment such as the assessment performed by the FEMA Expert Panel. Once issues associated with the biotechnology production process are resolved, safety assessment becomes routine and similar to that employed for conventionally produced flavor ingredients. The FEMA Expert Panel has adopted a GRAS assessment program which addresses these issues and can thereby assure the safety of flavor ingredients produced using genetically modified organisms.

		If	
Describe the product and characterize it chemically, then proceed to answer the following series of questions:			
		Yes No Go to	
1.	Is the product currently approved for use in foods?	2	Develop specifications and safety evaluation and go to 3 or reject
2.	Does the product meet existing specifications for identity and purity?	3	5
3.	Are the specifications adequate to ensure the absence and control of toxic constituents	4	5
4.	Do the intended or reasonably expected conditions of use of the product result in a pattern of intake that is supported by the safety database?	Accept	Accept with use limitations or do safety evaluation
5.	Do the consitutents pose no safety concern?	Revise specifica- tions and go to 4	6
6.	Can the undesired constituents be removed by processing?	Remove and go to 4	Safety evaluation; revise specifications and go to 4

Note: This figure was adapted from Ref. 26

Figure 1. Decision tree for the GRAS assessment of flavor ingredients derived from genetically modified microorganisms - safety evaluation

		If Yes	If No
		Proceed to	
1.	Does the microbe end up in food?	2	4
2.	Is the organism free of transferable antibiotic resistance genes?	4	3
3.	Does the resistance gene code for resistance to a substance used in control of disease agents in human or veterinary medicine?	Table 1, part D	4
4.	Are the vectors characterized and free of attributes that would render them unsafe for constructing microorganisms to be used to produce food-grade products?	5	Table 1, part D
5.	Does the DNA insert code for a substance safe for use in food?	6	Table 1, part D
6.	Is the microbe free of DNA from an intermediate host which could code for a toxic product?	Table 1, part A	Table 1, part D

Note: This figure was adapted from Ref. 26.

Figure 2. Decision tree for the GRAS assessment of flavor ingredients derived from genetically modified microorganisms - genetic modification issues.

Table I. Guidelines for Determining the Safety of Food and Flavor Ingredients Derived from Microorganisms[a]

A. Decision tree   If yes   If no
                                                                                    Proceed to
   1. Is the test material free of antibiotics[b]?   A2   D

   2. a. For bacteria and yeast:

         i. Is the test material free of toxins[c] known to be
            produced by other strains of the same species?   A3   D

         ii. If there are no known toxins[c,d] produced by
             other strains of the species, is the no-observable
             -effect level (NOEL) in a single oral challenge
             sufficiently high to ensure safety[e-g]?   B   D

      b. For molds, is the test material free of detectable
         levels of aflatoxin Bl, ochratoxin A, sterigmatocystin,
         T-2 toxin, zearalenone, and any other toxins known
         to be produced by strains of the same species[h]?   C   D

   3. Is the NOEL in short-term feeding studies sufficiently
      high to ensure safety[e-g]?   ACCEPT   D

B. Special considerations for certain yeasts and bacteria:

   1. If the source culture is a well-known, widely distributed, nonpathogenic yeast, e.g., certain species of the genus *Saccharomyces*, or if it belongs to a bacterial species that is well characterized, commonly present in foods, has a history of safe use in food ingredient manufacture, and has never been implicated in foodborne disease (e.g., *Bacillus coagulans, Bacillus licheniformis, Micrococcus lysodeikticus,* and *Bacillus subtilis*) the test material can be ACCEPTED at this point.

   2. Test material from other bacteria and yeasts must be considered under part A.3.

C. Special considerations for certain molds:

   1. If the source culture is well characterized, commonly present in food, has a history of safe use in food ingredient manufacture, and has never been implicated in foodborne intoxication or disease (e.g., *Aspergillus oryzae, Aspergillus niger,* and *Rhizopus oryzae*) the test material can be ACCEPTED at this point.

Table I. *Continued*

    2. Test material from all other species of molds must be considered under part, A.3.

D.   Disposition of materials that fail any decision tree requirements: A negative answer to question 1,2, or 3 signifies the presence of an undesirable substance and the material is not acceptable for use in food. If the undesirable substance can be removed, the purified material must be passed through the system again beginning at the noint of the original negative answer.

This table was adapted from Ref 26.

[a] These guidelines are intended for crude culture extracts, for whole cultures, and for concentrated enzyme or other microbially derived fractions which, when diluted, become preparations suitable for marketing.

[b] As determined by appropriate methods.

[c] For the purposes of these guidelines, the term toxin refers to a substance which is regarded by experts as a cause of food poisoning, intoxication, or illness when ingested. Examples are staphylococcal enterotoxins, botulinal neurotoxins, and mycotoxins.

[d] Certain cultures in this category are acceptable on the basis of single acute oral toxicity test, as explained in part B. 1. Cultures that fall under part B.2 can go directly to part A.3 without an acute oral toxicity test. This is permissible because the subchronic feeding specified in part A.3 is more rigorous and more meaningful than the acute oral toxicity test embodied in part A.2.a.ii.

[e] Expressed as mg/kg body wt and determined using appropriate animal species.

[f] Estimated mean consumption level is calculated from the sum of the intakes for each food category in which the material is expected to be used. An example of such determination is (USDA mean portion size) X (Market Research Corporation of American eating frequency for the entire population) X (the usual level of use expressed as total organic solids (TOS) for microbial preparation in question). TOS is defined as the sum of the organic compounds, excluding diluents, combined in the final microbral preparation.

[g] The term sufficiently high refers to appropriate multiples of the estimated mean human consumption level. For flavor ingredients the NOEL should be at least 100 times the estimated mean human consumption level.

[h] As determined by appropriate methods.

Acknowledgments

The authors thank Barbara West for help in preparing the manuscript for publication.

Literature Cited

1.   Shell, E.R. *Smithsonian,* **1987,** *6,* 79-88.
2.   IFT Expert Panel on Food Safety and Nutrition. Food flavors. *Food Technology,* **1989,** *43*(12), 99-106.
3.   Giese, J. *Food Technology.* **1994,** 48 (2), 106-116.
4.   Layman, P.L. *Chemical and Engineering News,* **1994,** *72*(23), 10-11.
5.   Hall, R.L. *Food Technology,* **1968,** *22,* 1388.
6.   Fischetti, F. In *CRC Handbook of Food Additives;* Editor, T.E. Furia. CRC Press, Boca Raton, Florida. *2,* 229-326.
7.   Gatfield, I.L. *Food Technology,* **1988,** *42*(10), 110-122.
8.   Romero, D.A. *Food Technology,* **1992,** *46*(11), 122-126.
9.   Nagodawithana, T. *Food Technology,* **1992,** *46*(11), 138-144.
10.   Bigelis, R. *Food Technology,* **1992,** *46*(11), 151-161.
11.   Janssens, L.; De Pooter, H.L.; Schamp, N.M.; Vandamme, E.J. *Process Biochemistry,* **1992,** *27,* 195-215.
12.   Armstrong, D.W.; Yamazaki, H. *Trends in Biotechnology,* **1986,** *4,* 264-268.
13.   Whitaker, R.J.; Evans, D.A. *Food Technology,* **1987,** *41*(9), 86-101.
14.   Knorr, D.; Beaumont, M.D.; Caster, C.S.; Dornenburg, H.; Gross, B.; Pandya, Y.; Romagnoli, L.G. *Food Technology,* **1990,** *44*(6), 71-79.
15.   Cheetham, P.S.J. *Trends in Biotechnology,* **1993,** *11*(11), 478-488.
16.   Flamm, E.L. *Biotechnology,* **1991,** *9,* 349-351.
17.   Food and Drug Administration. Direct food substances affirmed as generally recognized as safe: chymosin enzyme preparation derived from *Escherichia coli* K-12. 55 *Fed. Reg.* 10932 (23 March **1990**).
18.   Food and Drug Administration. Direct food substances affirmed as generally recognized as safe: chymosin enzyme preparation derived from genetically modified *Kluyveromyces marxianus* (Hansen) Van Der Walt variety Lactis (Dombrowski) Johannsen et Van Der Walt. 57 *Fed. Reg.* 6476 (25 February **1992**).
19.   Food and Drug Administration. Direct food substances affirmed as generally recognized as safe: chymosin enzyme preparation derived from genetically modified *Aspergillus niger.* 58 *Fed. Reg* . 27197 (7 May **1993**).
20.   Hodgson, J. *Biotechnology,* **1994,** *12*(8), 789-790.
21.   Food and Drug Administration. Statement of policy: foods derived from new plant varieties. 57 *Fed. Reg.* 22984 (29 May **1992**).
22.   Oser, B.L.; Ford, R.A. *Food Technology,* **1991,** *45*(11), 84-97.

23. Food and Drug Administration. Food flavoring: additional flavoring substances in GRAS review - extension of effective date for compliance with bulk flavor ingredient declaration requirements. 44 *Fed. Reg.* 71460 (11 December **1979**).
24. Woods, L.A.; Doull, J. *Regulatory Toxicology and Pharmacology*, **1991**, *14*, 48-58.
25. Oser, B.L.; Hall, R.L. *Food and Chemical Toxicology*, **1977**, *15*, 457-466.
26. International Food Biotechnology Council (IFBC). Biotechnologies and food: assuring the safety of foods produced by genetic modification. *Regulatory Toxicology and Pharmacology*, **1990**, *12*(Supp.), S1-S196.
27. Smith, R.L.; Ford, R.A. *Food Technology*, **1993**, *47*(6), 104-117.
28. Miller, H.I.; Gunary, D. *Science*, **1993**, *262*, 1500-1501.

RECEIVED June 7, 1995

PLANTS

Chapter 7

Determination of the Safety of Genetically Engineered Crops

Keith Redenbaugh, William Hiatt, Belinda Martineau, and Donald Emlay

Calgene, Inc., 1920 Fifth Street, Davis, CA 95616

The safety of FLAVR SAVR™ tomatoes was shown by a thorough evaluation and demonstration of substantial equivalence to non-genetically engineered tomato fruit. The FLAVR SAVR tomato does not differ from other tomato varieties except for the specific, intended effects of the inserted genes. All data support this conclusion, including molecular analyses, biochemical analyses, nutritional levels, horticultural traits, genetic analyses, field trial results, and plant pest risk evaluation. No data indicate or suggest any safety risk. Calgene also conducted a thorough review and analysis of the use of the kan^r gene and gene product, APH(3')II, for use as a selectable marker in tomatoes, cotton, and oilseed rape. The data generated concluded that APH(3')II is not a toxin or allergen, that the kan^r gene is highly unlikely to move from the plant genome into microorganisms via horizontal gene transfer, that if such transfer could occur the impact would be minimal, and that APH(3')II in transgenic plants will not compromise antibiotic use in humans or animals. On May 17, 1994, the FDA completed its evaluation of the FLAVR SAVR tomato and the use of APH(3')II, concluding that the tomato "is as safe as tomatoes bred by conventional means" (1) and "that the use of aminoglycoside 3'-phosphotransferase II is safe for use as a processing aid in the development of new varieties of tomato, oilseed rape, and cotton intended for food use" (2).

Calgene began discussions with the U.S. Food and Drug Administration (FDA) in February 1989. Based on these discussions, we submitted two extensive data packages to the FDA on the FLAVR SAVR tomato and the kan^r gene as "requests for advisory opinion," responded to nine requests for additional data, and consulted frequently with the FDA either in person or by telephone between 1989 and 1994. During this period, we also conducted eight field trials under permit from the U.S. Department of Agriculture, Animal and Plant Health Inspection Service (USDA APHIS). The evaluation of the FLAVR SAVR tomato is summarized in Table I.

0097–6156/95/0605–0072$12.00/0

Table I.   Consultations and Submissions by Calgene to the FDA and USDA

Date of Submission or Approval	Agency	Subject/Results	In response to
First meeting on 2/2/89	FDA	Discussion of safety of FLAVR SAVR tomato and selectable markers	
4/6/89	USDA	Approval to conduct field trial with FLAVR SAVR tomato in Yolo Co., CA	7 CFR 340
2/12/90	USDA	Approval to conduct field trial with FLAVR SAVR tomato in Ruskin, FL	7 CFR 340
3/19/90	USDA	Approval to conduct field trial with FLAVR SAVR tomato in Winters, CA	7 CFR 340
10/16/90	USDA	Approval to conduct field trial with FLAVR SAVR tomato in Indio, CA	7 CFR 340
11/26/90	FDA	kan^r gene and APH(3')II safety document submitted	Consultations with agency
5/1/91- 7/30/91	FDA	Only 43 public comments received on kan^r document	FDA requests public comments
5/22/91	USDA	Approval to conduct field trial with FLAVR SAVR tomato in Solano Co., CA	7 CFR 340
7/11/91	USDA	Approval to conduct field trial with FLAVR SAVR tomato in Manteca, CA	7 CFR 340
8/12/91	FDA	FLAVR SAVR tomato safety document submitted	Consultations with agency
12/17/91	USDA	Approval to conduct field trial with FLAVR SAVR tomato in Indio, CA	7 CFR 340
5/20/92	USDA	Approval to conduct field trial with FLAVR SAVR tomato in Central Valley, CA	7 CFR 340
5/31/92	FDA	kan^r gene used in human therapy, APH(3')II is not toxic or allergenic and is degraded in the gut	Consultations with agency
5/31/92	USDA	Petition to deregulate FLAVR SAVR tomato submitted	Consultations with agency and 7 CFR 340
7/14/92- 8/28/92	USDA	Only 24 public comments received on FLAVR SAVR tomato document	USDA requests public comments
10/19/92	USDA	Determination that FLAVR SAVR tomato is no longer a regulated article: 1) Approval letter of October 16, 1992 2) Interpretive Ruling, Fed. Reg. 57:47608	7 CFR 340
10/30/92	FDA	APH(3')II will not compromise antibiotic therapy	Consultations with agency
1/4/93	FDA	kan^r document converted to Food Additive Petition for APH(3')II	Consultations with agency

Continued on next page

Table I. *Continued*

Date of Submission or Approval	Agency	Subject/Results	In response to
3/1/93	FDA	Intubation studies showed no pleiotropic effects; vitamin levels as expected; inserted DNA stable; quality assurance standards proposed	Agency letters of 2/2/92 and 1/8/93
6/3/93	FDA	Revised environmental assessment (EA) for APH(3')II	Agency letter of 5/12/93
6/4/93	FDA	Only T-DNA transferred into tomato	Agency letter of 4/6/93
7/16/93-8/16/93	FDA	No public comments submitted on APH(3')II EA	FDA requests public comments
8/19/93	FDA	Neomycin in animal feed not inactivated	Consultations with agency
9/2/93	FDA	Safety studies lesions not test article related	Agency letter of 6/29/93
9/3/93	FDA	Tomatine levels equivalent	Agency letter of 6/8/93
10/1/93	FDA	Additional data on glycoalkaloids showed equivalence	Telephone consultation on 9/17/93
5/17/94	FDA	Approval of the FLAVR SAVR Tomato and APH(3')II 1) Advisory opinion letter of May 17, 1994 2) Letter Concluding Consultation, Fed. Reg. 59:26647 3) Secondary Direct Food Additive, APH(3')II, Fed. Reg. 59:26700	Calgene's request for safety determination

The objective of this paper is to provide a chronological description of the data and information requested by the federal agencies to determine that the FLAVR SAVR tomato was as safe as any other tomato variety. At the time Calgene began its discussions with the USDA and FDA, neither agency had policies in place for commercialization of transgenic crops. It was not clear, for example, that the FDA would use their postmarket authority to regulate foods produced using biotechnology. Since the FDA did not initially know how they would regulate such foods, our submissions were voluntary "requests for advisory opinions." Consequently, considerable interaction occurred as the process and types of information essential for safety assessment of transgenic crops were established. The end result was not only safety conclusions for the FLAVR SAVR tomato by both agencies, but also the establishment of specific, rational policies for commercialization ("deregulation"). These policies do not dictate which tests or experiments must be conducted in a safety evaluation, but provide guidance to companies and organizations. The USDA now has a petition process for a "determination of nonregulated status" for transgenic crops. FDA has established a consultation process to identify if any safety concerns exist for a new food or feed product and is using its postmarket authority for regulation.

Our approach to demonstrating the safety of the tomato followed the scientific principles that are the basis of the FDA 1992 Policy (3), the International Food Biotechnology Council report (4), and the Joint FAO/WHO Consultation (5). All of the data supporting the safety of the tomato and its status as food have been publicly available through the FDA (6, 7), journal articles (8, 9, 10), and a CRC Press book (11).

The extensive data generated on the FLAVR SAVR tomato demonstrated that it is as safe as any other tomatoes. To accomplish this we asked three basic questions.

1) Do we know what we inserted into the tomato?   The answer was yes.
2) Do we know the functions of the DNA we inserted?   The answer was yes.
3) Is the FLAVR SAVR tomato otherwise the same as any other tomato except for the intended changes?   The answer was yes.

These three basic questions are discussed in this paper.

Identity of FLAVR SAVR Tomatoes

FLAVR SAVR tomatoes contain a well characterized gene locus inserted into the tomato chromosome, which results in reduced levels of one tomato protein, polygalacturonase, and very small quantities of the APH(3')II protein. The FLAVR SAVR™ gene is an antisense polygalacturonase gene which causes a reduction in the production of the tomato enzyme polygalacturonase (PG). PG is an enzyme which degrades pectin, a natural constituent of the tomato fruit cell wall. The kanamycin resistance selectable marker gene (or kan^r gene) encodes the enzyme APH(3')II, which is aminoglycoside 3'-phosphotransferase II protein, also known as NPTII or neomycin phosphotransferase II.   APH(3')II phosphorylates two antibiotics, kanamycin and neomycin, and thus prevents them from inhibiting protein synthesis in plant cells. Because APH(3')II makes plant cells resistant to these antibiotics, we are able to select transformed cells in the laboratory.

The FLAVR SAVR gene transcribes an antisense (or reverse) RNA which, in theory, binds to the messenger RNA transcribed by the endogenous PG gene (which is labeled the "softening gene" in Figure 1). PG RNA is naturally transcribed into PG enzyme. One explanation of the mode of antisense inhibition is that these two RNAs tightly bind together as illustrated in Figure 1, and this greatly reduces the amount of the PG mRNA resulting in very low levels of PG (12).

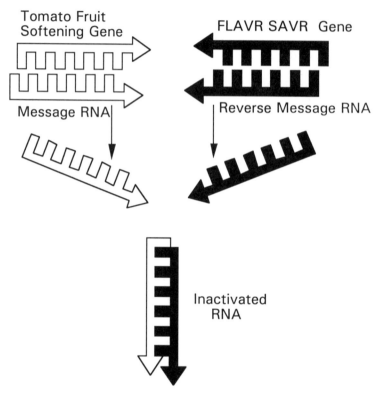

Figure 1. How FLAVR SAVR gene works. Reprinted with permission from Elsevier Trends Journals.

The purpose of developing FLAVR SAVR tomatoes is to improve the fruit taste by allowing large-scale commercial production of vine-ripe tomatoes. Currently, because of fruit softening during transport and storage, most fresh market tomatoes are picked at a green stage and gassed with ethylene to develop the red color. FLAVR SAVR tomatoes, because of the reduction in PG and delay in softening, can be left on the vine longer to develop better taste and flavor.

Question 1. Do We Know What We Inserted into the Tomato?

Yes, the process of plant genetic engineering is precise.   For Calgene's transformation work, we started with seven well characterized binary vector plasmids (13). All seven have the same plasmid backbone from pCGN1532 and each has right and left T-DNA border regions and the *kan*[r] selectable marker gene.   We fully characterized the DNA between the T-DNA borders in the 7 binary vectors.   For producing FLAVR SAVR tomatoes, we inserted the FLAVR SAVR gene into one of the 7 binary vectors, pCGN1547, to create the pCGN1436 plasmid.   The sources of the inserted genes are known: the antisense PG or FLAVR SAVR gene comes from tomato and the *kan*[r] gene comes from transposon Tn5 from *E. coli* K12 (6).

Insertion of the FLAVR SAVR and *kan*[r] genes into tomato was done using the standard *Agrobacterium tumefaciens*-mediated transformation system.

Southern hybridization analysis (14) was used in three ways to determine the structure of the inserted DNA: 1) we used restriction enzyme/probe combinations to identify the unique borders between the inserted DNA and tomato DNA in individual transformed plants, 2) we determined the number of inserted genes by comparing the intensity of the probe for the endogenous PG gene with the probe for the FLAVR SAVR gene, and 3) we used restriction enzyme/probe combinations to demonstrate that the *kan*[r] gene was physically linked to the FLAVR SAVR gene.   The information from these analyses was used to determine the structures of the inserted DNA in FLAVR SAVR tomato varieties.   Data for a number of tomato lines were submitted to the FDA.

Considerably more molecular analyses were done (2, 6) to provide the following information:

- The characterized DNA from pCGN1436, including the FLAVR SAVR and *kan*[r] genes, has the same structure in transformed plants as in the original plasmid.
- The number of copies of the inserted DNA were determined.
- The levels of PG and APH(3')II were determined.
- The inserted DNA behaved like any other tomato gene, segregating in a Mendelian fashion.

Question 2. Do We Know the Functions of the DNA We Inserted?

Yes, the intended effect of the *kan*[r] gene product APH(3')II is to provide resistance of plant cells to kanamycin in laboratory tissue culture growth medium. This occurred as predicted and we were able to select transformed cells. Regeneration on a medium containing kanamycin occurred only in transgenic cells.   Without this selection process, it would be difficult to select transgenic cells from non-transgenic ones, since the frequency of transformation is low.

The intended effect of the FLAVR SAVR gene was a significant and heritable reduction in PG enzyme activity.   Table II summarizes the measured levels of PG in a number of FLAVR SAVR tomatoes (6):

Table II.   **PG Enzyme Activities in FLAVR SAVR Tomatoes**

	T_1 generation	T_4	T_5
CR3 control	1.14-1.35[a]	1.17	0.95-1.2
CR3 1436	0.19-0.68	0.18-0.31	0.09-0.22

[a]PG data is reported as units of enzyme activity in red ripe fruit.

The result of reduced PG enzyme activity is a delay in fruit softening so that the tomatoes can remain on the vine for a longer time before picking and still survive transport to market.

Question 3. Is the FLAVR SAVR Tomato Otherwise the Same as Any Other Tomato Except for the Intended Changes?

Yes, we addressed the key nutritional components of tomato, vitamins A and C, monitored the naturally occurring toxicants, tomatine and solanine, and found that there were no significant changes. Although there are numerous other nutritional components of tomato that have been identified and measured, only vitamins A and C contribute significantly to the diet.

Nutrients. There was no variation in the major vitamins, protein amount and minerals as compared with non-transgenic tomato controls and natural ranges (6). Natural ranges in whole foods are very broad because of variations in growing conditions and differences among plant varieties. For example, vitamins A and C fall within the normal range (Table III). These components were also measured in FLAVR SAVR tomatoes after 4 generations and found to still be within the expected ranges. The most critical nutrients, vitamins A and C, have been measured in tomatoes from a number of transformation events and found not to differ from controls and the expected ranges.

Table III.   **Ranges of Nutrients (in 100 g fruit)**

Constituent	Normal range	Transgenics	Controls
Protein	0.85 g (.015 se)	0.75-1.14	0.53-1.05
Vitamin A	192-1667 IU	330-1600	420-2200
Thiamin	16-80 μg	38-72	39-64
Riboflavin	20-78 μg	24-36	24-36
Vitamin B6	50-150 μg	86-150	10-140
Vitamin C	8.4-59 mg	15.3-29.2	12.3-29.2
Niacin	0.3-0.85 mg	0.43-0.70	0.43-0.76
Calcium	4.0-21 mg	9-13	10-12
Magnesium	5.2-20.4 mg	7-12	9-13
Phosphorus	7.7-53 mg	25-37	29-38
Sodium	1.2-32.7 mg	2-5	2-3

In addition, we measured the levels of vitamins A and C in FLAVR SAVR and non-transgenic lines during the period that the fruit is palatable after harvest (i.e. throughout various stages of ripening) and found the amounts to be within the normal ranges, as expected.

Glycoalkaloids. Tomatoes naturally contain the glycoalkaloids tomatine and solanine. We compared the levels of these compounds in FLAVR SAVR and non-transgenic tomatoes and found no differences in tomatine amounts between FLAVR SAVR tomatoes and non-transgenic tomatoes for either green or ripe fruit (Table IV). These levels were within the normal range of tomatine in tomato fruit. The limit of detection was 0.25 mg tomatine per 100 g fresh weight of fruit. Also, as predicted, there was no detectable solanine in ripe fruit.

Table IV.   Tomatine Levels

Fruit Stage	Transgenics	Controls
Green	0 - 8.79 mg/100 g fwt	0 - 6.48
Red	0 - 1.09[a]	0 - 2.31[b]

[a]Only 1 of 38 fruit had detectable tomatine
[b]Only 4 of 60 had detectable tomatine

Field Trials. Data from eight field trials conducted under permits from USDA APHIS showed that FLAVR SAVR tomatoes were agronomically and horticulturally comparable to other tomato varieties. During these trials, the following characteristics were observed and measured to show that there were no changes in agronomic and horticultural traits:.

- Seed germination rate and frequency
- Flowering
- Outcrossing characteristics
- Pollination
- Fruit set
- Fruit color and size
- Fruit pH and acidity
- Seed production
- Plant size and form
- Harvest date
- Yield

Also, based on the 8 field trials, it was determined that there would not be any changes in growing practices compared to growing traditional vine-ripe tomatoes.

The USDA APHIS reviewed Calgene's data from the 8 field trials and determined on October 19, 1992 that the FLAVR SAVR tomato did not pose an environmental risk. They concluded that FLAVR SAVR tomatoes:

• Exhibit no plant pathogenic properties
• Are no more likely to become a weed than non-engineered parental varieties
• Are unlikely to increase the weediness potential for any other cultivated plant or native wild species with which the tomatoes can interbreed
• Do not cause damage to processed agricultural commodities
• Are unlikely to harm other organisms, such as bees, that are beneficial to agriculture

Potential for Allergenicity. APH(3')II was shown not to have homology with known allergens. It was compared with the known allergens in the GenBank, EMBL, PIR 29 and Swiss-Prot 19 databases. APH(3')II does not have the characteristics of an allergen:

• It is not glycosylated. The molecular weight of APH(3')II isolated from engineered plants is unchanged from the molecular weight of the bacterial protein, as demonstrated by SDS-polyacrylamide gel electrophoresis. This is not surprising because APH(3')II does not contain the necessary eukaryotic signal peptide sequences for transport into the endoplasmic reticulum (ER), which is where glycosylation reactions take place. Also, the protein does not contain the required amino acid sequences (asparagine-X-serine/threonine).
• APH(3')II is heat labile.
• It is present at very low concentrations, less than 35 μg per average tomato fruit.
• It is readily degraded under simulated gastric and intestinal conditions.

Potential for Toxicity. APH(3')II is present in very low levels and, like most proteins, is easily and quickly digested. APH(3')II was compared with known toxins in the GenBank, EMBL, PIR 29 and Swiss-Prot 19 databases and shown not to have homology with known toxins.

These conclusions on toxicity and allergenicity of the APH(3')II protein are supported by a recent publication on APH(3')II (15).

Potential for Compromising Antibiotics Therapy. We used a standard activity assay and found that kanamycin is not inactivated by APH(3')II under normal gastric conditions. Using tomato extracts in simulated normal or neutralized gastric conditions, we found that over the course of four hours there was no significant inactivation of kanamycin.

Neomycin and kanamycin are only used in two human antibiotic therapies, bowel surgery and hepatic encephalopathy. Because APH(3')II is rapidly degraded under gastric conditions, there is essentially no risk that an oral dose of neomycin or kanamycin could be compromised due to ingestion of transgenic fruits and vegetables containing APH(3')II.

However, we estimated the level of potential inactivation should the following unlikely conditions all occur simultaneously:

- 95th percentile consumption, at a single sitting, of specific fruits or vegetables high in ATP content.
- Calculations based on Pao et al. (16) survey of a three-day consumption period.
- Stoichiometric reaction of 100% of the ATP in ingested food with orally administered neomycin (this is highly unlikely).
- Administration of neomycin simultaneously with consumption of a transgenic food containing APH(3')II and with other fruits or vegetables rich in ATP.
- Presence of intact, functional APH(3')II enzyme, which requires a buffered stomach environment (pH 7).
- Stability of ATP in the stomach environment.

Even if all these conditions were met simultaneously, the potential reduction of antibiotic activity would be 1.2% of a 1 g dose of neomycin for tomatoes (Table V).

Table V.   Potential Reduction of Antibiotic Activity

Food	Consumption (g)[a]	ATP content μmol/100g	ATP consumption (μmol)	Potential inactivation % of 1 g dose[b]
Tomatoes	182	6.5	12	1.2
Carrots	122	7.5	9	0.9
Bananas	136	8	12	1.1
Oranges	228	6.5	15	1.5
Grapefruit	268	4.7	13	1.3
Spinach	205	5.8	12	1.1
Potatoes	280	2.4 - 12	7 - 34	0.7 - 3.5

[a]Consumption data are 95th percentile at a single sitting
[b]Neomycin amount in 1 g dose = 980 μmol

Therefore, the presence of the *kan*[r] gene and gene product APH(3')II do not result in any risk to the consumer.

Potential for Horizontal Gene Transfer.   Calgene developed a detailed, worst-case model to assess the potential of horizontal gene transfer (7). It should be noted that there is no known mechanism for plant to microbial DNA transfer; nevertheless, the model was designed to assume that such transfer could occur. Based on the results of the evaluations in this model, we concluded that:

- There will be no significant increase in exposure to kanamycin-resistant bacteria from consumption of the FLAVR SAVR tomato: at most 1 new *kan*[r] bacterial cell

will be produced for every 750 billion that are already present in the human gastrointestinal tract, and
- There will be no significant increase in kanamycin-resistant bacteria in agricultural soils: at most 1 new kan^r bacterial cell will be produced for every 10 million that are already present in soil.

Therefore, the kan^r gene is highly unlikely to move from the plant genome into microorganisms via horizontal gene transfer, and if such transfer could occur the impact would be minimal.

Conclusions of FDA Food Advisory Committee (FAC)

FDA held a public meeting on April 6-8, 1994, to discuss the safety issues of their 1992 Policy (3) using the FLAVR SAVR tomato as an example. The FAC concluded that all safety issues had been addressed for the FLAVR SAVR tomato (2):

- "The committee members generally expressed the view that the approach used by FDA to evaluate the safety of the tomato, including the safety of the kan^r gene, was appropriate and that all relevant scientific questions had been adequately addressed."
- "Transfer of the kan^r gene consumed as a component of tomatoes to microorganisms in the GI tract was highly unlikely."
- "The potential for transfer of the kan^r gene from plants to microorganisms in the environment is highly unlikely."
- If transfer could take place, "it would not cause a significant environmental impact."

FDA Conclusions on the FLAVR SAVR Tomato and kan^r Gene

On May 17, 1994, the FDA completed its evaluation of the FLAVR SAVR tomato (Figure 2) and reached the following conclusions.
FLAVR SAVR tomato: After considerable review, the FDA concluded that "FLAVR SAVR™ tomatoes have not been significantly altered when compared to varieties of tomatoes with a history of safe use" (17). In essence, then, the FLAVR SAVR tomato is just like any other tomato. There were no negative conclusions on the FLAVR SAVR tomato.
In its consultation letter and Memorandum of May 17, 1994 to Calgene (1, 17), the FDA stated that:

- The FLAVR SAVR tomato "is as safe as tomatoes bred by conventional means."
- "FDA has not found it necessary to require special labeling for FLAVR SAVR since it maintains the essential characteristics of traditionally developed tomatoes."

Aminoglycoside 3'-phosphotransferase II: Calgene requested that the FDA consider the use of the kan^r marker gene protein, aminoglycoside 3'-

DEPARTMENT OF HEALTH & HUMAN SERVICES   Public Health Service

_____

Food and Drug Administration
Washington DC 20204

Mr. Donald L. Emlay   MAY 1 7 1994
Director, Regulatory Affairs
Calgene, Inc.
1920 Fifth Street
Davis, CA 95616

               Re: FMF 526 and Docket No. 91A-0330

Dear Mr. Emlay:

This is in response to your request, dated August 12, 1991, for consultation with the
Food and Drug Administration (FDA) concerning FLAVR SAVR™ tomatoes. You
requested that FDA issue an advisory opinion under 21 CFR 10.85 concerning whether
FLAVR SAVR™ tomatoes are food and, therefore, subject to the same regulation as
other tomato varieties. This request is separate from your request for the evaluation of
safety of the aminoglycoside-3'-phosphotransferase II (APH(3')II) protein used for
selection of plant cells that incorporated the new genetic trait. Because the question
concerning APH(3')II is addressed in a separate rulemaking, this letter addresses only
aspects of FLAVR SAVR™ tomatoes other than APH(3')II.

In the *Federal Register* of May 29, 1992 (57 FR 22984), FDA issued a "Statement of
Policy: Foods Derived from New Plant Varieties", in which FDA advised that requests
for consultation with the agency should be made consistent with the principles outlined in
the policy. Therefore, we are treating your request as a consultation in accordance with
the May 1992 policy statement.

As noted in that statement (57 FR 22984 at 22990), FDA has rarely had the occasion to
review the regulatory status of foods derived from new plant varieties because these
foods have been widely accepted as safe. FDA regulations in 21 CFR 170.30(f) do,
however, provide for review of the regulatory status of certain substances of natural
biological origin in certain circumstances. Specifically, 21 CFR 170.30(f)(2) provides
for the review of the regulatory status of any substance of natural biological origin with
a history of safe use that has had "significant alteration of composition by breeding or
selection." Based on the information that Calgene has submitted concerning the FLAVR
SAVR™ tomato, we believe that this new variety has not been significantly altered
within the meaning of 21 CFR 170.30(f)(2), when compared to varieties of tomatoes
with a history of safe use.

Although the agency has concluded that the FLAVR SAVR™ tomato has not been
significantly altered, we emphasize that under the Federal Food, Drug, and Cosmetic
Act (the Act), it is Calgene's responsibility to ensure that foods that the firm markets are
safe, wholesome, and in compliance with all provisions of the Act.

                    Sincerely yours,

                    Alan M. Rulis. Ph.D.
                    Acting Director
                    Office of Premarket Approval
                    Center for Food Safety
                      and Applied Nutrition

Figure 2. May 17, 1994 FDA Letter (1).

phosphotransferase II (APH(3')II) as an indirect food additive for use in tomato, oilseed rape, and cotton. In a detailed report (2), FDA reached a number of conclusions that the marker gene protein was safe for use as a processing aid in these crops. There were no negative conclusions on the use of APH(3')II in these three crops:

- "FDA has concluded that the use of aminoglycoside 3'-phosphotransferase II is safe for use as a processing aid in the development of new varieties of tomato, oilseed rape, and cotton intended for food use."
- "No limits other than good manufacturing practice are needed to ensure the safety of the petitioned use of APH(3')II."
- "FDA has determined that there is no need to set a tolerance for the amount of APH(3')II that will be consumed because the agency knows of no reason why this protein would have any properties that would distinguish it toxicologically from any other phosphorylating enzymes in the food supply."
- "FDA agrees with Calgene that the characteristics of APH(3')II do not raise a safety concern."
- "APH(3')II is not known to be toxic."
- "APH(3')II has been shown to be rapidly degraded under simulated gastric conditions."
- "The estimated dietary exposure to APH(3')II is very low (480 µg APH(3')II per person per day, or 0.16 ppm."
- "Processed products that contain tomatoes with the kan^r gene are unlikely to contain any enzymatically active APH(3')II."
- "The biological activity of APH(3')II is destroyed during gastric and intestinal phases of digestion."
- "Any active APH(3')II that might remain would not significantly inactivate kanamycin or neomycin in the gut."
- "FDA concludes that the presence of APH(3')II in food will not compromise the therapeutic use of orally administered kanamycin or neomycin."
- "Oils derived from transgenic cottonseed and rapeseed modified using the kan^r gene would not be expected to contain active or inactive APH(3')II."
- "There was no significant difference with respect to neomycin stability between medicated cottonseed and rapeseed meals prepared from transgenic cottonseed and rapeseed containing APH(3')II."
- "Transgenic strains of cottonseed and rapeseed containing APH(3')II have no apparent untoward effect regarding the stability of neomycin and that the therapeutic efficacy of neomycin in animal feed will not be affected."
- "FDA does not consider those substances that are inherent components of food to be ingredients that must be disclosed in the food's label. A genetic substance introduced into a plant by breeding becomes an inherent part of the plant as well as of all foods derived from the plant. Consistent with FDA's general approach on ingredient labeling, the agency has not treated as an ingredient a new constituent of a plant introduced by breeding, regardless of the method used to develop the new plant variety. Accordingly, FDA has determined that neither the kan^r gene

nor APH(3')II is an ingredient that, under section 403 (i) of the act, must be individually identified in labels of foods containing them."

Kanamycin resistance marker gene: The FDA also reviewed the *kan*^{r} gene and the potential for transfer to other organisms (2), concluding that:

- There would be "no increase in kanamycin-resistant soil microorganisms."
- "Even if such transfer could occur, the rate at which it could occur is such that it would not result in a detectable increase over the existing background population of kanamycin-resistant bacteria."
- "The use of the *kan*^{r} gene does not pose safety concerns in terms of increase in the population of antibiotic-resistant pathogens due to the potential for horizontal transfer of the gene."
- "Introduction of the *kan*^{r} gene will not confer a competitive advantage upon a plant receiving it."
- "Transfer of the *kan*^{r} gene to other crops or related weeds will have no significant adverse environmental effects."
- "Probability of transfer of the *kan*^{r} gene to gut microflora is remote."
- "The *kan*^{r} gene will not compromise the efficacy of antibiotic treatment."
- "There is no evidence that free DNA containing the *kan*^{r} gene, even if present, can transform cells lining the GI tract."

However, FDA did note that their conclusions were only for tomato, oilseed rape, and cotton (2):

- "The potential transfer of the *kan*^{r} gene, as well as other antibiotic resistance marker genes, from crops to microorganisms should be evaluated on a case-by-case basis."
- "Approval of Calgene's petition would not mean that developers could use the *kan*^{r} gene in crops other than those identified in the petition."

Conclusions

The FDA has been very consistent in asserting how it will regulate, not only foods derived using biotechnology techniques, but all foods. FDA has repeatedly stated that the postmarket authority under section 402 (a) (1) of the Federal Food, Drug and Cosmetic Act will "continue to be the primary legal tool for ensuring the safety of whole foods derived from genetically modified plants" (2), and that guidance to industry, universities and other organizations can be found in the 1992 Policy (3).

These actions by the FDA validated the 1992 Policy (3), allowed commercialization of the FLAVR SAVR tomato, and provided clear guidelines on the type of safety assessment FDA felt was appropriate in evaluation of new plant varieties.

References

(1) FDA. Letter to Calgene on FLAVR SAVR tomato. May 17, 1994. (FMF 526 and Docket No. 91A-0330).

(2) Federal. Register 59:26700-26711, 1994. Secondary Direct Food Additives Permitted in Food for Human Consumption; Food Additives Permitted in Feed and Drinking Water of Animals; Aminoglycoside 3'-Phosphotransferase II.

(3) Federal Register 578:22984-23003, 1992. Statement of Policy: Foods Derived from New Plant Varieties; Notice.

(4) International Food Biotechnology Council. Biotechnologies and food: assuring the safety of foods produced by genetic modification. Regul. Toxic. Pharm. 12 (3) Part 2 of 2 Parts: S1-S196, 1990.

(5) Joint FAO/WHO Consultation. Strategies for assessing the safety of foods produced by biotechnology. Geneva, World Health Organization, 1991, 59 pages.

(6) Redenbaugh, K.; Hiatt, W.; Martineau, B.; Kramer, M.; Sheehy, R.; Sanders, R.; Houck, C.; Walker, D.; Emlay, D. Request for Food and Drug Administration Advisory Opinion. FLAVR SAVR™ Tomato: Status as Food. Docket No. 91A-0330, 1991.

(7) Emlay, D.; Hiatt, W.; Houck, C.; Malyj, L.; Kramer, M.; Kiser, J.; Knauf, V.; Lindemann, J.; Martineau, B.; McBride, K.; Mitten, D.; Redenbaugh, K. Request for Food and Drug Administration Advisory Opinion. kan^r Gene: Safety and Use in the Production of Genetically Engineered Plants. Docket No. 93F-0232, 1990.

(8) Redenbaugh, K.; Berner, T.; Emlay, D.; Frankos, B.; Hiatt, W.; Houck, C.; Kramer, M.; Malyj, L.; Martineau, B.; Rachman, N.; Rudenko, L.; Sanders, R.; Sheehy, R.; Wixtrom, R. Regulatory issues for commercialization of tomatoes with an antisense polygalacturonase gene. In Vitro Cell. Dev. Biol. 1993, 29P, 17-26.

(9) Redenbaugh, K.; Hiatt, W.; Martineau, B.; Emlay, D. Regulatory assessment of the FLAVR SAVR tomato. Trends Food Sci. Tech. 1994, 5, 105-110.

(10) Redenbaugh, K.; Hiatt, W.; Martineau, B.; Lindemann, J.; Emlay, D. Aminoglycoside 3'-phosphotransferase II (APH(3')II): Review of its safety and use in the production of genetically engineered plants. Food Biotech. 1994 (in press).

(11) Redenbaugh, K.; Hiatt, W.; Martineau, B.; Kramer, M.; Sheehy, R.; Sanders, R.; Houck, C.; Emlay, D. Safety Assessment of Genetically-Engineered Fruits and Vegetables: A Case Study the FLAVR SAVR™ Tomatoes. CRC Press: Boca Raton, FL, 1992, 267 pages.

(12) Sheehy, R.; Kramer, M.; Hiatt, W. Reduction of polygalacturonase activity in tomato fruit by antisense RNA. Proc. Natl. Acad. Sci. USA. 1988, 85, 8805-8809.

(13) McBride, K.; Summerfelt, K. Improved binary vectors for Agrobacterium-mediated plant transformation. Plant Mol. Biol. 1990, 14, 269-276.

(14) Southern, E. Detection of specific sequences among DNA fragments separated by gel electrophoresis. *J. Mol. Biol.* **1975**, *98*, 503-517.

(15) Fuchs, R.; Ream, J.; Hammond, B.; Naylor, M.; Leimgruber, R.; Berberich, S. Safety assessment of the neomycin phosphotransferase II (NPTII) protein. *Biotech.* **1993**, *11*, 1543-1546.

(16) Pao, E.; Fleming, K.; Guenther, P.; Mickle, S. Foods Commonly Eaten by Individuals: Amount per Day and per Eating Occasion. Home Economics Research Report Number 44, USDA Consumer Nutrition Center, Washington, D.C, 1982.

(17) Federal Register 59:26647-26648, 1994. Calgene, Inc.; Availability of Letter Concluding Consultation (Docket No. 91A-0330).

RECEIVED July 14, 1995

Chapter 8

Exotic Germ Plasm or Engineered Genes

Comparison of Genetic Strategies To Improve Fruit Quality

Alan B. Bennett, Roger Chetelat, and Ellen Klann[1]

Mann Laboratory, Department of Vegetable Crops,
University of California, Davis, CA 95616

Genetic improvement of crops is a well-established practice which has greatly advanced agricultural productivity and food quality. Recently, molecular genetic approaches have been utilized which enhance the efficiency and precision of classical plant breeding approaches and provide the opportunity to introduce completely novel genes into transgenic plants. In the context of food safety, it is useful to consider the genetic consequences of a classical or genetic strategy to modify a particular plant trait. Such a direct comparison was possible in tomato, where a trait which modifies carbohydrate composition was introgressed from a wild tomato species or engineered by introduction of an antisense transgene. From this comparison, it was possible to evaluate the precision of the genetic change in relation to the size and position of introduced DNA. Both classical and molecular genetic approaches introduce a degree of uncertainty in the final genetic makeup, but the source of this uncertainty is quite different in each case. It is reasonable that analysis of the safety of genetically engineered food products be evaluated relative to that of traditional approaches that achieve the same end.

Genetic modification of both plants and animals has been the basis of agricultural development. This process was initiated by the earliest domestication of crops from wild ancestral species and has continued for over 10,000 years. The most recent manifestations of this continuing process have been the green revolution, where focused efforts lead to particularly rapid modifications in plant performance, and the application of recombinant DNA technologies which promises

[1]Current address: Biology Department, University of Massachusetts, Amherst, MA 01003

to expand the genetic resource base beyond what is accessible through classical genetic hybridization. It is appropriate to consider how different approaches to genetic modification of crops may effect the safety of the resulting food product. One way to approach this consideration is to examine specific examples of genetic modification by either classical plant breeding methods or by the application of recombinant DNA techniques and to assess the extent and predictability of resulting genetic changes. This approach cannot determine the safety of a given product but provides the basis to rationally consider the potential sources of unexpected alterations in the food product that could have consequences for safety. There is one example where a single trait influencing tomato sugar composition has been the target of a dual genetic approach and provides the basis of making such a direct comparison.

Carbohydrate Composition of Tomato

Tomato fruit typically accumulate the hexose sugars, glucose and fructose throughout development (1). Sugars are the major component of fruit total soluble solids (TSS) and this parameter is the single most important contributor to the processing quality of tomatoes (2). Because processed tomato products are concentrated by evaporation to a fixed level of TSS increases in sugar concentration in the raw fruit can lead to large increases in processed product yield and to reduced processing costs. Estimates indicate that increases in TSS of 0.1% have a value of approximately $7 M to the California tomato processing industry alone. In addition, sugar content is an important determinant of flavor and therefore contributes to the quality of fresh tomatoes as well.

Efforts to improve tomato TSS levels have recognized that the trait is polygenic (3), with recent studies suggesting that at least three chromosomal regions contributed to elevated TSS in one experimental population (4,5,6). Because there is a negative association between yield and TSS levels, selection for high yielding tomato varieties has resulted in a trend towards reduced TSS levels (7). In addition to strictly genetic approaches to elevate TSS levels, there have been ongoing efforts to identify biochemical determinants of high sugar levels and use that information to direct selection strategies (8,9,10,11,12,13,14). This latter approach assumes that improvement of a biochemical process that is a component of a complex trait can contribute incrementally to enhancement of the overall trait (15). A biochemical component that has been suggested to contribute to elevated sugar concentration in tomato fruit is the modification of sugar composition by increasing sucrose accumulation (11,12,14,16). Because tomato fruit normally accumulate hexose sugars and very low amounts of sucrose, this change alters the sucrose/hexose ratio in favor of sucrose. The basis for suggesting that sucrose accumulation would contribute to elevated total sugar levels is based on the decreased osmotic contribution of sucrose relative to hexose, which predicts that sucrose-accumulating fruit would accumulate less water during development and maintain a higher sugar concentration than the corresponding hexose-accumulating fruit (13,14). To test this proposal, tomatoes have been genetically modified to accumulate sucrose by both classical genetic means and by genetic engineering. The modification of this single trait by both approaches provides a basis to directly compare the genetic consequences of each approach.

Plant Breeding and Sources of Genetic Variation

Plant breeding relies on classical genetic analysis and procedures to enhance crop characteristics. The hallmark of plant breeding is that the process selects beneficial traits from naturally occurring diversity. This imposes a significant limit, in that the process of plant breeding can only access the diversity that exists in the same or closely related species with which the crop plant can be sexually hybridized. This limitation is especially pronounced in tomato (*Lycopersicon esculentum*) which has very little genetic variation and so represents a shallow pool for selection of useful traits (17). The low level of genetic variation in tomato appears to have resulted from its history of domestication. The center of genetic diversity of tomato is in the Andean region of South America, but it was domesticated as a food crop in Mexico where it is likely that there were relatively few genetic stocks available for these initial selections. Tomato was then transported from Mexico to Europe where the genetic sources of modern tomato lines originated. This movement of tomato first to Mexico and then to Europe represented two bottlenecks that most likely reduced the genetic variation in tomato. Fortunately, there exist in the Andean region of South America, a number of wild relatives of tomato that possess extensive genetic variation. Even though these wild relatives are distinct species (*Lycopersicon chilense*, *L. hirsutum*, *L. chmielewskii*, *L. peruvianum*, *L. pinpinellifolium*, *L cheesmanii*, and *L. pennelii*) they can be hybridized with tomato, although in some cases only with great difficulty (17). The wild tomato relatives are themselves largely inedible and have poor horticultural traits such as small fruit size, but they represent a genetic source of resistance to a large number of diseases as well as a source of numerous fruit quality traits, such as high TSS.

Two species, *L. chmielewskii* and *L. hirsutum*, were identified to accumulate sucrose, rather than hexose as in domesticated tomato, *L. esculentum* (11,13,18). In addition to accumulating sucrose, both of these wild relatives of tomato also accumulate very high levels of total sugar, consistent with the idea that this trait could contribute to elevated total sugar concentration. Previous research had also demonstrated the feasibility of using *L. chmielewskii* and *L. hirsutum* as parents in breeding programs with *L. esculentum* (19,20). Collectively, these observations indicated that there was a source of gene(s) that could be accessed by classical genetic strategies to confer sucrose-accumulation in tomato.

Introgression of Sucrose Accumulation from a Wild Tomato Species

The trait of sucrose accumulation in *L. chmielewskii* was transferred to tomato, *L. esculentum*, by crossing individuals of the two species (14). The first progeny of a cross between these two parents all accumulated hexose sugars, indicating that the trait of sucrose accumulation was recessive. Subsequent generations were obtained by backcrossing to a *L. esculentum* parent in order to progressively increase the relative contribution of *L. esculentum* to the genome (Figure 1). In the early generations, the segregation ratios indicated that the trait of sucrose accumulation was controlled by a single recessive gene which was named *sucr* (16). This conclusion meant that the trait could be transferred in a relatively straightforward manner into *L. esculentum*.

Introgression of traits by classical genetic means is characterized by the transfer of large blocks of DNA that may encompass entire chromosomes or large segments of chromosomes. Because the *L. chmielewskii* genome encodes many undesirable traits, our goal was to transfer the minimum segment of the *L. chmielewskii* genome to *L esculentum* that conferred the sucrose accumulating trait. The use of molecular markers corresponding to the genome of the donor and recipient species provides the basis to achieve this goal, but requires that we first map the trait of sucrose accumulation to a specific chromosomal region. This was accomplished by screening a series of restriction fragment length polymorphism (RFLP) markers in two populations segregating for the trait of sucrose accumulation (Figure 2). The RFLP analysis indicated that the trait resided on chromosome three and was tightly linked to the DNA marker, TG102, providing a means to indirectly select for the trait of sucrose accumulation by selection for individual plants carrying the *L. chmielewskii* allele of TG102. This indirect selection allowed screening of seedlings in both F_1 and F_2 populations for the fruit-specific trait and greatly accelerated the process of introgression.

In addition to accelerating the backcross process, the identification of molecular markers linked to the trait of sucrose accumulation also provided the basis to select plants which carried the minimum segment of the *L. chmielewskii* genome required to confer the trait of sucrose accumulation. Figure 3 illustrates schematically the process by which molecular markers were used to specifically introgress the *sucr* locus of *L. chmielewskii* into *L. esculentum*. Sucrose-accumulating individuals of a third generation backcross family (BC_3) were identified and their detailed genotype of chromosome three determined by scoring for a number of RFLP markers along the chromosome. In the example shown (Figure 3) approximately one third of chromosome three was comprised on *L. chmielewskii* genome, and this region encompassed the TG102 locus which had been previously shown to be linked to *sucr*. This individual plant was backcrossed twice more to the recurrent *L. esculentum* parent and sucrose-accumulating progeny of the fifth generation backcross were scored for RFLP markers along chromosome three. As indicated in Figure 3, recombination events occurred at several positions which yielded individual progeny with reduced segments of the *L. chmielewskii* genome, that nevertheless encompassed the TG102 locus (21).

The use of molecular markers in the introgression of the *sucr* locus accelerated the rate of introgression by providing a basis for scoring individual plants in the seedling stage and for identifying heterozygotes in F_1 populations. In addition, the markers provided a basis to identify the chromosomal region carrying the *sucr* locus and to select plants carrying the minimum *L. chmielewskii* genomic fragment necessary to confer sucrose accumulation. As indicated above, introgression of traits by classical genetic means is characterized by the transfer of large blocks of DNA. Even using molecular markers to minimize the size of the introgressed fragment, we have determined that the introgressed *L. chmielewskii* fragment carrying the *sucr* locus is between 0.5 and 7 centiMorgans. Although there is no strict relationship between genetic and physical distances in the tomato

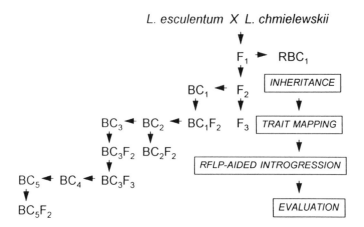

L. esculentum X L. chmielewskii

Figure 1: Diagram of a backcross strategy to introgress the trait of sucrose accumulation from *L. chmielewskii* to *L. esculentum*. The initial F_1 progeny accumulated hexose but sucrose-accumulating individuals were recovered in all subsequent F_2 populations.

BC_2F_2 Population:

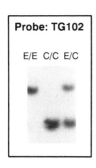

Probe: TG102

E/E C/C E/C

TG102	PHENOTYPE	
	Hexose	Sucrose
E/E	73	0
E/C	106	1
C/C	1	21

Reciprocal BC_1 Population:

TG102	PHENOTYPE	
	Hexose	Sucrose
E/C	133	0
C/C	0	120

Figure 2: Restriction fragment length polymorphism (RFLP) analysis indicates that the DNA probe, TG102, is tightly linked to the trait of sucrose accumulation. Based on different sizes of the TG102 DNA fragment from *L. chmielewskii* or *L. esculentum* (inset) the origin of the TG102 genomic fragment was assessed in two populations segregating for sucrose accumulation. In almost every case, sucrose-accumulating plants were homozygous for the *L. chmielewskii* allele of TG102, indicating that the trait is physically linked to this genomic fragment in the tomato genome.

genome, we estimate that this chromosome segment could comprise up to five megabases of DNA and could potentially encode hundreds of genes other than those corresponding to the *sucr* locus.

Tomato lines derived from marker-aided introgression of the *L. chmielewskii* TG102 locus accumulated sucrose to levels equivalent to the parent wild species but also exhibit negative horticultural traits, such as reduced fertility and scattered fruit set (22). The associated traits may be pleiotropic effects of the introgressed *sucr* gene or may result from the action of linked genes carried on the introgressed *L. chmielewskii* fragment.

Molecular Cloning of a Gene Responsible for Sucrose Accumulation

Cloning of the gene responsible for sucrose accumulation in tomato was initiated by identifying the biochemical basis of the trait (13,14,23). Analysis of a number of enzymes in tomato fruit that could reasonably contribute to elevated sucrose levels indicated that the level of a single enzyme, invertase, was greatly reduced in sucrose accumulating *L. chmielewskii* and in sucrose-accumulating backcross derivatives this wild species (14). Because invertase catalyzes the breakdown of sucrose to fructose and glucose, the finding of low invertase levels was consistent with its conferring the trait of sucrose-accumulation. Antibodies and cDNA probes that specifically react with invertase protein and mRNA were used to determine that the reduced invertase enzyme activity resulted from the absence of the invertase protein and mRNA in sucrose-accumulating fruit (14,23). This result strongly implicated the gene encoding invertase as being responsible for the trait of sucrose accumulation, with the *L. chmielewskii* invertase gene differing from the *L. esculentum* invertase gene by having no expression in fruit.

Having identified invertase as the likely biochemical basis of sucrose accumulation, it was possible to use what have become standard techniques for cloning its gene. Because the nucleotide sequence of a gene specifies the amino acid sequence of its corresponding protein, we purified the invertase protein and determined its partial amino acid sequence. The corresponding nucleotide sequence of its gene was thus deduced and used to synthesize short DNA sequences corresponding to parts of the predicted invertase gene sequence. This nucleotide probe was then used to screen libraries and to isolate and sequence the complete invertase gene (24).

Although biochemical evidence suggested that the invertase gene conferred the trait of sucrose accumulation, an alternative possibility was that a second gene, which controlled invertase expression in fruit, provided the genetic basis of the trait. To distinguish between these two possibilities, the chromosomal location of the invertase gene was mapped relative to the TG102 locus which we had previously shown to be tightly linked to the trait of sucrose accumulation. In a BC_5F_2 population, segregating for the trait of sucrose accumulation, alleles of TG102 and the invertase gene (TIV1) co-segregated, indicating that these two loci are very tightly linked in the tomato genome (Figure 4a). This tight linkage of TG102 and TIV1 is illustrated in Figure 4b where the two loci are shown to be coincident on chromosome three. Because we had previously shown that TG102 is tightly linked to *sucr*, this result strongly supports the conclusion that the

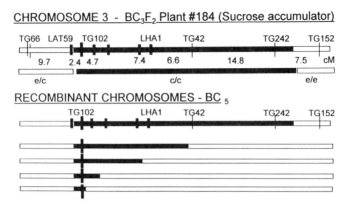

Figure 3: Marker aided introgression of the TG102 locus from *L. chmielewskii*. The genotype of chromosome 3 in a single individual sucrose-accumulating plant (# 184) in the BC₃F₂ population was determined (upper panel). The diagram indicates segments of the chromosome that were derived from *L. esculentum* (e), *L. chmielewskii* (c) or from both parents (e/c). A population of BC₅ individuals derived from plant #184 were similarly analyzed and a number of individuals identified with greatly reduced *L. chmielewskii* segments of chromosome 3 (lower Panel).

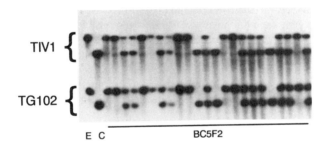

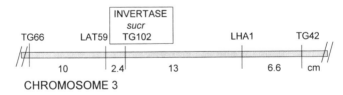

Figure 4: Genomic Southern blot of *Eco*RI-digested DNA isolated from *L. esculentum* (E), *L. chmielewskii* (C) and individuals of a BC₅F₂ population segregating for sucrose accumulation. The blot was probed with TG102 or TIV1. The *L. esculentum* and *L. chmielewskii* alleles of TG102 and TIV1 co-segregate in all cases. Based on the cosegregation of TG102 and TIV1, and the linkage of *sucr* to TG102, we infer that TG102, TIV1 and *sucr* map to genetically identical locations of tomato chromosome three, as indicated on the schematic diagram of molecular markers on the chromosome (lower diagram). Upper panel reprinted with permission from ref. 16.

invertase gene itself is the genetic basis of sucrose accumulation in *L. chmielewskii* and its derivatives.

Genetic Engineering of Sucrose Accumulation in Tomato

Identification of the invertase gene, and the absence of its expression in fruit, as the genetic determinant of sucrose accumulation suggested an alternative strategy to engineer this trait in transgenic plants by down-regulating the endogenous *L. esculentum* invertase gene. The down-regulation of endogenous genes in plants has been accomplished by the expression of antisense transgenes (25,26) or by the phenomenon of co-suppression where expression of sense transgenes also resulted in suppression of endogenous gene expression (27). We constructed chimeric antisense invertase genes for expression in tomato in order to test whether direct suppression of endogenous invertase gene expression would be sufficient to confer the trait of sucrose accumulation. In these experiments, two types of antisense genes were constructed, one type being regulated by a constitutive promoter (35S) and the other type being regulated by a fruit ripening-specific promoter (28), to guard against the potentially deleterious effects of suppressing invertase gene expression in all plant parts (Figure 5). The chimeric gene is comprised of approximately 2.0 kb of the invertase coding sequence (cDNA) and was cloned into an Agrobacterium-based transformation vector, pBIN 19 (29). Both the chimeric gene fragments and the vector have been sequenced in their entirety (30).

In all cases, the constitutive (35S) promoter was more effective in reducing expression of the endogenous invertase gene and had no deleterious effects on plant growth and development when plants were grown either in the greenhouse or field. When sugar composition of transgenic fruit expressing the antisense gene was analyzed, several transgenic lines accumulated sucrose to levels comparable to *L. chmielewskii* (Klann and Bennett, unpublished data). This confirmed that low invertase gene expression was sufficient to confer the trait of sucrose accumulation. Further analysis of these transgenic lines indicates that they also accumulate a higher concentration of total sugars and do not suffer from negative horticultural traits such as the low fertility observed in lines derived from crosses with *L. chmielweskii*.

Genetic Modifications Associated with Sucrose Accumulation

In this paper we have described a single genetically-determined biochemical modification that alters the ratio of soluble sugars in tomato fruit to favor sucrose accumulation. This trait is potentially favorable from the perspective of enhancing total soluble sugar concentration. The unique feature of this trait with respect to the assessment of the safety of genetically modified foods is that the same trait has been conferred by both classical plant breeding as well as by genetic engineering. Because the same trait was conferred in both cases (e.g. the product of each change was substantially identical), it is possible to focus on the genetic changes associated with the process conferring the phenotypic change.

In considering safety aspects of genetic modifications there are chiefly two concerns:

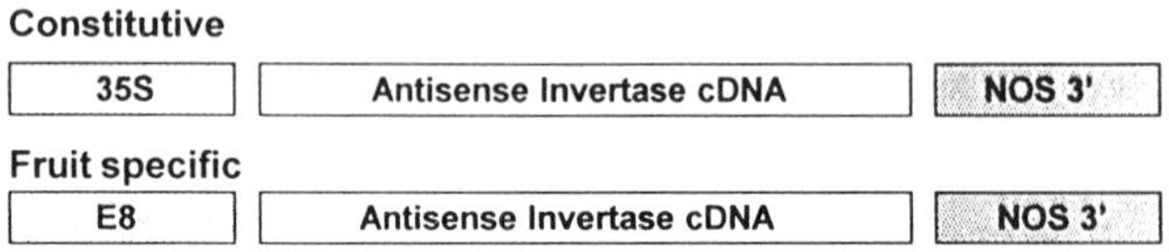

Constitutive

| 35S | Antisense Invertase cDNA | NOS 3' |

Fruit specific

| E8 | Antisense Invertase cDNA | NOS 3' |

Figure 5: Schematic diagram of chimeric antisense genes to inhibit invertase gene expression in transgenic tomato. The 35S promoter allows for antisense gene expression in most plant tissues, whereas the E8 promoter specifies expression in ripening fruit. The 3' terminator of nopaline synthase was included in each chimeric gene construct.

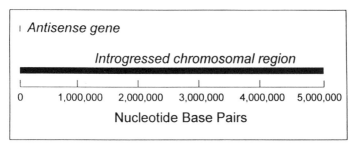

Figure 6: Size comparison of introduced DNA fragments in sucrose accumulating plants derived by transfer of an antisense gene (upper bar; approximately 2 kb) or by introgression of an *L. chmielewskii* genomic fragment of chromosome three (lower bar; up to 5 Mb).

1.   That the added gene products are themselves safe and do not confer toxicity or allergenicity to the modified food product. To assess the safety of the modified food product in this context, it is important to have a high degree of certainty of the identity and sequence of the added gene.

2.   That the genetic modification did not produce unexpected results either by pleiotropy, where a gene of known function indirectly effects other plant traits, or by insertional mutagenesis, where the added gene may integrate at the site of an endogenous functional gene and disrupt its activity. Safety assessment in this context suggests that it would be important to have a high degree of certainty of the chromosomal location of the introduced gene. Collectively, the certainty of genetic changes that are introduced when plant traits are modified by either plant breeding or genetic engineering can be characterized by knowledge of the introduced gene identity (sequence) and the chromosomal location of its insertion.

Using the best techniques available, the trait of sucrose accumulation was conferred by introgression of a relatively large chromosomal fragment from *L. chmielewskii* or by the direct transfer of a very small antisense gene (Figure 6). In the case of the antisense gene, its sequence and the sequence of all of the transferred genes was precisely known and so can be readily evaluated for potentially toxic or allergenic gene products. In the case of the introgressed *L. chmielewskii* genomic fragment, although we know that the invertase gene resides on this fragment, the transferred genetic material is sufficient to encode up to a thousand genes. Because the sequence of this fragment is not precisely known, it can not be readily evaluated for potentially toxic or allergenic gene products.

Conversely, the chromosomal location of the introgressed *L. chmielewskii* genomic fragment is precisely known because this fragment always integrates at its native site on chromosome 3, whereas the antisense gene inserts randomly into the recipient genome (31). Thus, the certainty of the chromosomal location of the introgressed fragment is high, and that of the antisense gene is low. In each method of genetic modification; plant breeding characterized by the transfer of large undefined genomic fragments at precise a chromosomal location or genetic engineering characterized by the introduction of precisely defined genetic elements at random sites; a degree of uncertainty over the genetic makeup of the resulting food product is introduced.

Conclusions

Crop improvement by an iterative process of evaluation and selection is the foundation of agriculture. This practice has been greatly advanced by the application of genetic principles and most recently by the use of DNA-based molecular markers that increase both the precision and speed of introgression of traits from related species. This practice of plant breeding is generally regarded as safe, and indeed we have thousands of years of experience with relatively few introductions of unsafe food as a result of plant breeding. Nevertheless, such examples of the introduction of unsafe foods do exist, most likely do to the introduction of undesirable genes linked to the introgressed genomic fragments. In spite of the advances in plant breeding that have dramatically increased its power, breeding is ultimately limited by its access only to traits that reside in plant species that are closely related to the crop species and can be sexually hybridized. Genetic engineering overcomes this limiting barrier and provides the possibility of accessing traits from essentially any other organism. We have relatively less experience with this technology as compared to plant breeding, yet it is reasonable to evaluate the comparative genetic changes introduced by each approach as a means to ask whether one technology introduces greater inherent uncertainty in the genetic makeup of the resulting food.

The example of sucrose-accumulation in tomato provides a unique example where a single trait has been conferred by both classical and molecular genetic approaches. From this comparison, it was possible to evaluate the precision of the genetic change in relation to the size and chromosomal location of the introduced DNA. Both approaches introduce a degree of uncertainty in the final genetic makeup. Classical genetic introgression introduces uncertainty in the identity of the introgressed genomic fragment and genetic engineering introduces uncertainty in the chromosomal location of the introduced gene. In either case the potential exists for introducing either toxic or allergenic gene products or having unanticipated effects resulting from pleiotropy or positional effects. Thus, it is appropriate that the safety of genetically engineered food products be evaluated relative to that of traditional, and widely accepted, plant breeding approaches that achieve similar ends.

Acknowledgements

This research was supported by Binational Agricultural Research and Development Fund grant number US-1872-90.

Literature Cited

1.   Davies, J.N.; Kempton, R.J. *J. Sci. Food Agric.* **1975**, *26*, 1103-1110.
2.   Stevens M.A.; Kader, A.A.; Albright-Holton, M.; Algazi, M. *J. Am. Soc. Hort. Sci.* **1977**, *114*, 695-699.
3.   Ibaria, E.A.; Lambeth, V.N. *J. Am. Soc. Hort. Sci.* **1969**, *94*, 496-498.
4.   Osborn, T.C.; Alexander, D.C.; Fobes, J.S. *Theor. Appl. Genet.* **1987**, *73*, 350-356.
5.   Paterson, A.H.; Lander, E.S.; Hewitt, J.D.; Petersen, S.; Lincoln, S.; Tanksley, S.D. *Nature* **1988**, *355*, 721-726.
6.   Tanksley, S.D.; Hewitt, J.D. *Theor. Appl. Genet.* **1988**, *75*, 811-823.
7.   Stevens, M.A. *Plant Breeding Rev.* **1986**, *4*, 274-310.
8.   Dinar, M.; Stevens, M.A. *J. Am. Soc. Hort. Sci.* **1981**, *106*, 415-418.
9.   Hewitt, J.D.; Marush, M. *J. Am. Soc. Hort. Sci.* **1986**, *111*, 241-245.
10.   Ho, L.C. *Ann. Rev. Plant Physiol. & Plant Molec. Biol.* **1988**, *39*, 355-378.
11.   Miron, D.; Schaffer, A.A. *Plant Physiol.* **1991**, *95*, 623-627.
12.   Stommel, J.R. *Plant Physiol.* **1992**, *99*, 324-328.
13.   Yelle, S.; Hewitt, J.D.; Robinson, N.L.; Damon, S.; Bennett, A.B. *Plant Physiol.* **1988**, *87*, 737-740.

14. Yelle, S.; Chetelat, R.T.; Dorais, M., DeVerna, J.; Bennett, A.B. *Plant Physiol.* **1991**, *95*, 1026-1035.
15. Bennett, A.B.; Chetelat, R.; Klann, E; Lashbrook, C.; Martin, R.; Gilchrist, D. *Trans. Malay Soc. Plant Physiol.* **1992**, *3*, 200-209.
16. Chetelat. R.T.; Klann. E.; DeVerna. J.W.; Yelle. S.; Bennett. A.B. *Plant J.* **1993**, *4*, 643-650.
17. Rick, C.M.; Yoder, J.I. *Ann. Rev. Genet.* **1993**, *22*, 281 -300.
18. Davies, J.N. *Nature* **1966**, *209*, 640-641.
19. Rick, C.M. *Hilgardia* **1974**, *42*, 493-510.
20. Rick, C.M.; Kesicki, E.; Fobes, J.F.; Holle, M. *Theor. Appl. Genet.* **1976**, *47*, 55-68.
21. Chetelat, R.T.; DeVerna, J.W.; Bennett, A.B. *Theor. Appl. Genet.* **1995a**, (submitted).
22. Chetelat, R.T.; DeVerna, J.W.; Bennett, A.B. *Theor. Appl. Genet,* **1995b**, (submitted).
23. Klann, E.M.; Chetelat, R.T.; Bennett, A.B. *Plant Physiol.* **1993**, *103*, 863-870.
24. Klann, E.M.; Yelle, S.; Bennett, A.B. *Plant Physiol.* **1992**, *99*, 351-353.
25. Sheehy R.E.; Kramer, M.; Hiatt, W.R. *Proc. Natl. Acad. Sci. USA* **1988**, *85*, 8805-8809.
26. Smith, C.J.S.; Watson, C.F.; Ray, J.; Bird, C.R., Morris, P.C.; Schuch, W.; Grierson, D. *Nature* **1988**, *334*, 724-726.
27. Napoli, C.; Lemieux, C.; Jorgensen, R. *Plant Cell* **1990**, *2*, 279-289.
28. Deikman J.; Fischer R.L. *EMBO* **1988**, *7*, 3315-3320.
29. Bevan M. *Nucl. Acids Res.* **1984**, *12*, 8711-8721.
30. Fray, R.G.; Wallace, A.D.; Grierson, D. *Pl. Molec. Biol.* **1994**, *25*, 339-342.

RECEIVED June 7, 1995

Chapter 9

The Potential for Allergenicity in Transgenic Foods

Oscar L. Frick

Department of Pediatrics, University of California,
San Francisco, CA 94143

Anaphylactic shock and allergies may occur in sensitized man and animals even in minute quantities after being ingested or inhaled. Therefore, such exposure of a food-sensitive individual to an allergenic substance in a transgenic food might constitute a safety risk. The likelihood that a transferred gene is also the allergen in that food is very small, but it has occurred. The pathophysiology of allergic sensitization and reactions is reviewed. Many allergens in common foods have been identified. Genes from plant and animal sources are being transferred to other food crops to ensure improved shelf-life, taste, nutritional value, and resistance. Brazilnut 2S protein put into soy added more sulfur containing amino acids, but also transferred the potent Brazilnut allergen to soy, as evidenced by the transgenic soy binding to nut allergic persons' IgE antibodies in vitro. Therefore, safety of the new transgenic food in food-allergic subjects must be addressed. This concern, however, should not hinder development of such potentially valuable new foods.

As this is primarily a chemical-oriented audience, I have been asked to describe the biology of allergy and review some aspects of potential allergens that might be transferred to genetically altered foods. This is an area of concern to consumers, food producers and to the Food and Drug Administration.

Two years ago, I was asked to evaluate allergy in an 18-month old girl who broke out in hives after she ate some peanut butter cookies. Allergy skin tests showed a very strong reaction to peanut. During her pregnancy, the mother craved peanut butter and probably sensitized her infant in utero. My advice was to avoid all peanut-containing foods, read labels, and I gave her an epinephrine Epi-Pen and an anti-histamine to administer in an emergency. One year ago, the mother took the child on an airplane trip to San Diego. About 30 minutes into the flight after the hostess served drinks and packets of peanuts to the other passengers and they opened their packets, the child began sneezing, broke out in hives and began

0097–6156/95/0605–0100$12.00/0

choking and wheezing. Mother recognized that peanut fumes in the air caused a reaction in the baby, and she gave her an epinephrine injection and an anti-histamine. Fortunately, the child responded within minutes and stopped choking, breathed easier and the remaining reaction subsided during the rest of the flight. This is an example of how minute an exposure to a potent allergen could cause anaphylactic shock with possible disastrous consequences.

Anaphylaxis is a shock syndrome caused by a massive generalized release of histamine and other allergic mediators from even minute exposures to an antigen. Food proteins are common causes of anaphylaxis, especially peanut, tree-nuts, eggs, fish and shell-fish. Other common causes are stinging insect venoms and drugs, like penicillin.

Biotechnology applied to plant and animal food sources is a further extension of the Green Revolution that will be needed to feed the world's exponential explosion in population growth, especially in the next centuries. Biotech is already here on a commercial scale in that chymosin, the bio-engineered substitute for calf-stomach rennin that clots casein, is already present in 50% of the cheeses available on the U.S. market. This year BST (bovine somatotropic hormone) was approved for increasing cow's milk production (about 20% per cow) so that farmers require less pasture land and feed to produce the same amount of milk. This is a boon to dairy farmers, especially in the developing world.

In plant food crops, this biotechnology developed in the past decade; 1982 was the first successful genetic transformation in a tobacco plant (1). First, genes for desirable agronomic traits were identified and isolated from the whole plant DNA. An appropriate promoter sequence was added to these desired cloned genes to allow expression to make a novel plant variety. Field trials then followed to ascertain stability, performance and safety of the new plant variety.

Currently plants (1) have been developed with: 1) genes that confer resistance to various insects - either genes of bacterial origin that code for potent insecticidal protein (Bt protein from *Bacillus thuringiensis*) or plant genes that code for proteinase inhibitors. 2) genes coding for herbicide-insensitive version of plant target enzymes or genes coding for herbicide-detoxifying enzymes ("Round-up", Monsanto). 3) genes that protect against viral infection. 4) genes coding for enzymes that control ripening process where gene expression can be decreased by using "antisense" technology to limit rate of ripening (Flavr-Savr tomato, Calgene). 5) some herbicide and antibiotic resistance genes can be used as marker genes to distinguish between transferred and non-transferred cells early on by coexpression of the desired gene and the marker. Coming soon are genes that confer: cold resistance, drought resistance, male sterility, nutritional or sensory enhancement of food plants. An ultimate may be a banana into which anti-viral and anti-bacterial disease vaccines have been introduced which would permit massive vaccinations of tropical populations by their eating such transgenic bananas.

However, there are some potential risks in bioengineered foods that concern the public; one of these is the possible transfer of allergens in transgenic foods. An anti-freeze gene from flounder fish has been transferred to such frost-sensitive fruits as strawberry and citrus. Is a fish-allergic person at risk of anaphylaxis when he eats a transgenic strawberry or citrus? This is of major concern for consumers, food processors, and the Food and Drug Administration. The FDA in 1992 issued guidelines for food safety aspects of new plant varieties with a strong provision about potential allergenicity transfer.

What is an Allergy?

It is an aberrant or hyperimmune response to non-noxious proteins or glyco-proteins in foods or environment. The body makes diverse types of immune reponses to fight off microorganism invaders - bacteria, viruses, fungi and altered or foreign tissues - tumors and grafts; such immune responses are expected to be beneficial to the body.

In the usual helpful immune response, the invader bacterium or virus is attacked by white blood cells, such as neutrophils and monocytes, latter become activated macrophages. Invader components, especially proteins and carbohydrates, are antigens which are ingested and processed by macrophages. Fragments of the antigen are then brought to the macrophage surface in conjunction with the macrophage's own major histocompatibility antigen (MHC) for presentation to other immune cells, the T-lymphocyte, especially of the helper-types, sometimes of a suppressor type (2). Recently (3), it has been recognized that there are two types of T-helper lymphocytes Th-1 and Th-2. Such T-lymphocytes when activated by antigen presented by macrophages, release cytokines - chemicals that act on other cells or have biologic activities - these now number in the scores.

In the usual protective immune reactions, the microbial antigen taken up and presented by the macrophage with its MHC interacts with a complex T-cell receptor on the T-helper-1 (Th-1) lymphocyte; these release an array of cytokines - especially interleukin-2 (IL-2) and interferon gamma (IFN-γ) which activate B-lymphocytes specific for that antigen to differentiate into an antibody-producing plasma cell producing immunoglobulin antibodies of either the IgM, IgG or IgA class. Such antibodies react in tissues or blood stream with the invader microorganism which it agglutinates or activates plasma factors, such as complement proteins to lyse and destroy the microorganism.

Alternatively, the Th-1 lymphocyte activates through its cytokines - IL-2 and IFN-γ, other cytotoxic lymphocytes and macrophages which elaborate other cytokines that kill the invader, e.g., tubercle bacillus, or tumor cell, or foreign transplanted tissue cell, so-called cellular immunity.

For immune defense against parasite worms, a second T-helper (Th-2) lymphocyte reacts with a macrophage that presents its processed parasite antigen with its MHC. The activated Th-2 lymphocyte releases a different Th-1 set of cytokines - primarily interleukins-4 and -5 (IL-4, IL-5) which induce parasite-specific B-lymphocytes to differentiate into plasma cells producing a different immunoglobulin E (IgE) antibody. IgE antibodies react with different sets of inflammatory cells - eosinophils, basophils and mast cells that release potent chemical toxins locally to destroy the parasite.

This latter same immune system of Th-2 lymphocytes, IL-4, IL-5, B-lymphocytes producing IgE antibodies and mast cells, eosinophils and basophils cause allergic reactions.

In temperate climates and with better hygienic measures in food-handling and water-purification, endemic parasites have been controlled and there appears to be less need for the IgE antibody anti-parasite defense. Therefore, in temperate climates, most families have forgotten how to make Th-2 induced IgE immune reactions. Yet about one-third of temperate climate families still make substantial amounts of IgE antibodies to non-noxious glycoprotein antigens called allergens; therefore, a genetic predisposition to allergies. American-born children of immigrant families from parasite-endemic tropical regions have 3 times higher IgE levels than American Caucasian children (4), even though they have never been exposed to parasites - such children are prone to develop severe allergic diseases, such as asthma and eczema. Allergies are increasing in incidence and severity around the globe, as public hygiene improves, but environmental pollution increases. Over the past decade, increased IgE levels and allergies have been associated with viral infections (5) caused by crowding, and polyphenols in cigarette smoke (6) and in Diesel fumes (7). Also the 1974 Oil Crisis induced tighter building construction and wall-to-wall carpeting to save heating fuel, but also created a more favorable environment for propagation of house dust mites, *Dermatophagoides*, which are the world's leading cause of allergies.

I shall now return to the mechanism of the allergic reaction. If there is a genetic predisposition to allergy and conditions are right, the first time that person experiences contact with that allergen (plant pollen, animal, food) entering a natural portal (nose, bronchi, skin, GI tract) results in IgE antibody production. The newly formed IgE antibodies circulate briefly (1 day) in the blood and then enter the tissues to attach to or sensitize target cells - usually mucosal and submucosal mast cells, also blood basophils and eosinophils which have specific receptors for IgE antibodies; such a person is allergically sensitized.

Upon the subject's next and subsequent contacts with the allergen, the absorbed allergen finds its way via the circulation to the specific antibodies on the mast cells. Because allergens are multivalent, they react with two or more IgE antibody molecules bringing them and their receptors into proximity. This leads to a series of intracellular activation steps with mobilization of intracellular calcium and influx of extracellular calcium which causes fusion of the mast cell granule membranes and extrusion and emptying on the mast cell's surface. These granules contain potent preformed inflammatory mediators - histamine, heparin, tryptase, chemokines, and also activation causes membrane phospholipid breakdown to platelet activating factor (PAF) and arachidonic acid which metabolizes to newly-formed mediators, prostacyclines and leukotrienes (8).

The pharmacologic effects of these mediators are listed (9). Among them are histamines which cause vascular dilation and leakage of plasma into tissues (edema-hives, hayfever), smooth muscle contraction and glandular secretion (asthma). Leukotrienes LTC4, LTD4, LTE4 are potent sustained smooth muscle constrictors (asthma, diarrhea) and vasodilators and LTB4 is a potent chemoattractant for other inflammatory cells (neutrophils, eosinophils, basophils, macrophages). PAF is an extremely potent vasodilator (probably the main mediator of anaphylactic shock) and eosinophilotactant. These agents cause the classic immediate allergic reaction that occurs within minutes and subsides usually within an hour.

The allergy skin test (10) in which a drop of allergen solution placed on the skin is pricked into the skin results within minutes of vasodilation (redness), itching, and edema (wheal) due to the release of histamine from the allergic reaction. Formerly, it was thought that the immediate reaction was all that there was, and medications were developed to block or reverse effects of immediate reactions - anti-histamines and broncho-dilators, such as β-adrenergic agonists and theophylline.

However, a decade ago, a late-phase allergic reaction was recognized. If a very large immediate skin reaction occurred and subsided in one hour, subsequently at 6 to 8 hours later, there was a second larger brawny indurated skin reaction which subsides only after many hours to days.

Similarly, in many, but not all, asthmatic patients, a purposeful inhalation exposure to an allergen, e.g., dust mite extract, resulted in an immediate broncho-constriction, measured by a fall in a pulmonary function test, such as forced expiratory volume in 1 second (FEV1) or peak expiratory flow (PEF). This bronchospasm lasted about one hour and pulmonary tests returned to baseline. However, with no new exposure to allergen, 4-6 hours later a second, usually more profound, bronchoconstriction occurred which lasted many hours or even days. Tissue biopsies of the bronchi or bronchial lavage fluids revealed many inflammatory leukocytes, especially eosinophils and basophils in the site of the late-phase reaction. Eosinophils (11) have granules with highly basic proteins (pH > 10), such as major basic protein (MBP), eosinophil cationic protein (ECP), eosinophil peroxidase (EPO) which are released when the eosinophil is activated in inflammation. These inflammatory proteins remain in the allergic reaction-damaged tissue for days, even weeks, causing further tissue damage. Other inflammatory cells, such as basophils, macrophages, lymphocytes also contribute to tissue damage. Similar late-phase allergic reactions occur in the skin (atopic dermatitis or eczema) (12), nose (perennial allergic rhinitis with secondary sinusitis (13), eyes (chronic allergic conjunctivitis) (14), and most recently, we have shown such late-phase reactions in the stomach resulting from a food-allergic reaction (15). The recent recognition of late-phase allergic inflammation has radically changed the pharmacologic management of asthma and allergies in that now we use anti-inflam-matory drugs, such as topical corticosteroids, cromolyn and nedocromil, as the workhorse agents.

Allergic reactions to foods are most common in infants and pre-school children; foods cause up to 30% of allergies in such children. The mucosal surface protective secretory IgA class of antibodies develops slowly in children, reaching adult level by age 7. Secretory IgA dimers in the gastrointestinal tract react with and precipitate incompletely digested proteins and proteoses in foods (that is not broken down completely to amino acids). Such proteose-IgA antibody complexes are eliminated in the intestinal stream, thereby limiting the absorption of incompletely digested foods. With an underdeveloped secretory IgA system in infants and young children with a genetic propensity for allergy, some incompletely digested proteins do cross the intestinal mucosal barrier and react with a second defense line, Th2 and B-lymphocytes that form IgE antibodies to the food allergens. If one extrapolates that a formula-fed infant drinks the equivalent of 16 gallons of foreign protein, cow's milk, per day; it is quite obvious that some incompletely digested milk proteins do get through the intestinal barrier and sensitize the infant. It is a wonder that allergic sensitization does not happen even more often. As the protective secretory IgA antibody reaches maturity at 5-7 years of age, the incidence of food allergies decreases. The major food allergens (16) in children are cow's milk, soy, egg, wheat and peanut (ubiquitous use of peanut butter). In older children and adults, fish, shell-fish, tree nuts (Brazil, walnut, hazelnut), corn and tomato become more important food allergens. In adults, food allergies account for only about 1-2% of all allergic problems, usually manifested by anaphylactic shock which can be lethal, urticaria, gastro-intestinal and respiratory symptoms.

Transfer of genes from one animal or plant to another food product of animal or plant origin presents the potential problem of transferring an allergen from one species to another. Therefore, if a person allergic to the gene donor plant or animal eats a transgenic food with that allergen gene, is he likely to experience an allergic reaction? That is the concern of the consumer and the FDA. I think that the risk is extremely minute considering the number of genes in a plant or animal and the chance that the particular gene transferred for its beneficial effect is also the same gene that is an allergen. Much work is being done by allergists to isolate and identify the allergic epitopes, that is the chemical grouping in a protein that causes an allergy.

Allergens

Many of the allergens in food have been well characterized and sequenced and even crystallized for x-ray diffraction and NMR structure studies. Protein isolation methods, such as SDS-PAGE, monoclonal antibodies, allergen cloning and sequence analysis and production of recombinant allergens, and analysis of T-cell responses to allergens provided insights into regulation of IgE responses.

There are some major differences between allergy to inhaled allergens, like pollens, house dust mites, and animal danders, and ingested food allergens (17). Inhaled allergens are relative low molecular weight (10-50 kD) which become airborne on small particles (2-50 microns). Amounts inhaled are minute (1-10 μg/year) which causes IgE antibody formation by stimulating specific Th2-lymphocytes to elicit IL-4 and IgE. Normal individuals make little or no IgG antibodies to these inhaled allergens. In contrast, food allergens are ingested in large quantities - grams to centograms; all people make IgG antibodies to them and 0.1-1% of people make IgE antibodies - these latter are the allergics. Therefore, for food allergens, there appear to be both Th1-type response with IFN-γ and IgG production and a Th2-response with IL-4 and IgE production.

The functions of many food allergens have been identified: β-lactoglobulin is a retinol-binding protein, ovomucoid is a trypsin-inhibitor, codfish allergen is a parvalbumin calcium-binding protein, and shrimp allergen is a tropomyosin. Chapman (18) found a common allergen denominator among β-lactoglobulin, the major allergen in cow's milk, and insect proteins - Bla g 4 cockroach allergen, tobacco hornworm insecticyanin, butterfly bilin-binding protein - all of which are transporters and binders of small hydrophobic ligands called calycins. Although their sequence homology is only about 20%, they have 3 structure conserved regions (SCR) and have similar 3-dimensional structures with the SCR at one end in a cupshaped ligand-binding pocket. This region appears to be the allergen in both cockroach βLa g 4 and β-lactoglobulin. Furthermore, cow's milk β-lactoglobulin may be a strong allergen in children because it is lacking in human breast milk, thus, it is the most foreign protein in cow's milk to humans. In cow's milk, α-casein is a strong allergen, while some allergic children have IgE antibodies to β- and γ-casein, α-lactalbumin, BSA and BGG. α_{sl}-Casein binds strongly to surface receptors on mouse T- and B-lymphocytes, but this phenomenon has not been studied on human Th2-cells.

In the late 60s, Berrens (19) recognized that Maillard adducts, the amino-carbonyl reaction products between lactose and lysine in protein were strong allergens, not only in cow's milk, but also in tomato, kapok, and feather pillow stuffings. Progressive blockade of ε-amino groups of lysine side-chains in peptides of allergens with aldose sugars causes relative preponderance of contribution to net negative charge of acidic amino acids. This lysine-sugar appendage is one of the enolic forms II or III of its Amadori configuration of I, and adds to the acidity at positions 2 II or 3 III.

Bleumink and Berrens (20) made extracts from green oranges, unripe tomatoes, also fresh ripe red tomatoes from the same plant, and also ripe red tomatoes stored 14 days. Allergenic skin reactivity increased dramatically from unripe to ripe status due to Maillard non-enzymatic "browning" causing formation of N-glycosidic 1-amino deoxy ketone groups.

Egg is a common allergen. In a long prospective study of children born into allergic families, Roundtree et al. (21) found IgG antibodies to egg and milk from 3 months age on. About 30% (27/92) children developed IgE antibodies to egg associated with eczema and asthma in the first 2 years of life. Inhalant allergy to mites and pollen developed later.

Ovomucoid and ovalbumin are the most prominent allergens in egg. Ovomucoid is a heat-stable glycoprotein even in hard-boiled eggs and contains 20-25% carbohydrates. The allergenic epitope in ovalbumin has been identified as regions 323-339 and a synthetic peptide made with these amino acids is equally reactive with IgE of egg allergic patients (22). This was confirmed (23) by cyanogen bromide cleavage fragments of ovalbumin where residues 41-172 and 301-385 reacted with patients' IgE antibodies. Conceivably, one could put anti-sense DNA probes into hens to produce non-allergenic eggs. There are, however, other weaker allergens among the main 13 egg proteins, such as ovotransferrin, ovoinhibitor, and weakly lysozyme in egg white and livetin in egg yolk.

Legume allergens, especially peanut, are particularly strong, as I illustrated by my opening case presentation. Legumes have lectins that non-specifically stimulate lymphocytes; phytohemagglutinin (PHA) from jack beans is widely used. Such lectins causing adjuvant activity to a specific immune allergic stimulation might explain legumes' high allergenicity.

Peanut has two main storage proteins, arachin and conarachin. Barnett and Howden (24) found the major allergen to be a subfraction of these, a heat-stable concanavalin A-reactive glycoprotein of 65 kD with 2% sugar and removing sugar did not eliminate reactivity with IgE antibodies. Burks et al. (25) isolated a similar glycoprotein allergen (Ara h I) of 63.5 kD which did not bind concanavalin A and also a 17 kD glycoprotein (Ara h II) which cross-reacts with Ara h I, which could possibly be a fragment.

Recently, Burks et al. (25) with Dorian (26) studied T-cell responses to Ara h II in 9 peanut allergic asthmatics, and in 12 controls - 6 normal persons and 6 non-peanut allergic asthmatics. Lymphoproliferation with both Ara h II and a ubiquitous pan-sensitizer Candida mold occurred in all - and were significantly greater than to other allergens - ovalbumin, soy, and casein. Most striking, however, was that normals and asthmatics produced IFN-γ (the Th1 cytokine) to Ara h II, while the peanut allergics produced no IFN-γ. All groups made IFN-γ to Candida. In their assay, they were unable to detect IL-4. However, they induced cytokine gene transcripts with Ara h II for IL-4 in 2 peanut allergics (Th2-cytokine), but not for IFN-γ, while 2 normals made good transcripts for IFN-γ, but not IL-4; both groups made good transcripts for β-actin control. This study confirms the Th1 and Th2 responses in food allergy and suggests that the level of IFN-γ production in response to Ara h II may be an important factor in determining the development of peanut specific IgE response.

Soybean. There are at least 16 soybean proteins that react with IgE from soy-allergic patients. Kunitz soybean trypsin inhibitor (SBTI) is anaphylactogenic in some (27). Because SBTI is widely used in biochemical research, it can be an occupational hazard in sensitized personnel. IgE from persons allergic to both soy and peanuts bind large proteins (50-60 kD), while IgE from soy only, not peanut, allergics bind a 20 kD 2S-globulin (28). In soy-sensitive Japanese children with atopic dermatitis, a 30 kD 7S globulin is the major soy allergen (29). Furthermore, an 11S globulin, glycinin, was the allergen in 8 patients - 4 to a subunit, so in soybean, there doesn't appear to be one major allergen, but many (30).

Peas. A green pea albumin, stable on boiling, was reacted with IgE of pea-allergic patients. The purified allergen was 1.8 kD and had 30% sugars.

Castor bean allergen has been identified as a glutamine-rich albumin storage protein from endosperm and has a similar amino acid sequence as lima bean protease inhibitor.

Cereal allergens are related to grass pollens, because all cereals are cultivated grasses.

Rice is a major allergen in Japan, especially in children with atopic dermatitis. Matsuda et al. (31) found several 14-16 kD rice albumins to be the main allergens for which the complete amino acid sequence is known from cDNA nucleotide sequence. It is quite similar to α-amylase/trypsin inhibitor family from other cereals (wheat, barley) and legume (castor bean).

Wheat. Baker's asthma patients who inhale wheat flour during baking have IgEs that bind to several wheat proteins - albumin, wheat germ agglutinin, a concanavalin A-binding glycoprotein, and a trypsin inhibitor (32).

Fish. The first well-defined food allergen was codfish muscle allergen M (Gad c I), a parvalbumin which is a calcium chelator protein that regulated Ca^{2+} concentration in muscles (33). Gad c I is a 12 kD protein with 113 amino acids and one glucose. This or a similar protein appears in many food fishes, but not all, so fish-allergic patients are advised to avoid all fishes. However, our recent study showed some fish, especially the elasmobranch shark, had quite different allergens from bony fishes (34). Also some fish-sensitive patients can tolerate canned tuna and canned salmon because prolonged heating during processing denatures the fish allergens (35).

Crustacea include shrimp, prawns, crab, lobster, and crayfish. The major allergens in several shrimp species are heat stable 34-36 kD glycoproteins which share sequence homology with tropomyosins (36). These also have been identified in crab, prawns and crayfish and are stable even after boiling. There is considerable amino acid sequence homology of shrimp tropomyosin to those in beef, chicken and pork, but not, as yet, IgE-binding in meat-allergic persons. In San Francisco restaurants, a few tiny Bay shrimp are often put into green salads, so shrimp-sensitive patients must be wary of salads. I have had one patient who had anaphylaxis from shrimp-traces that were in a washed clean bowl in which a salad with shrimp had been tossed before.

Foods cross-reacting with inhalant allergens. Apples, peaches, and hazelnuts contain Bet v 1 (37), the main allergen in birch tree pollen (the "ragweed" of Scandinavia), so Northern climate birch hayfever patients often have an "oral allergy syndrome" with itching and swelling of tongue and buccal mucosa when they eat apples, peaches or hazelnuts, especially in Spring birch season.

Profilin (38), designated Bet v II, which controls actin polymerization in all eukaryotic cells occurs in all vegetable foods and plant pollens. Profilin is a common epitope in grass and mugwort pollens to which 20% of pollen-sensitive patients make IgE antibodies. Such IgEs also bind celery, carrots, apple and potato, but don't react with human profilin (39).

Finally, most plants express PR proteins when damaged or viral-infected and these may also be allergens. These PR proteins may be expressed when an apple or potato is peeled due to cell damage.

Transgenic foods

Returning now to biotechnology-engineered foods, I indicated that such foods are already on the American market in chymosin-treated cheeses and in BST-cow's milk.

This year, several tomato-products have been or will be approved and will appear in our markets soon. The Flavr-Savr™ tomato (Calgene) has an enhanced shelflife (1). Polygalacturonase degrades pectin in tomatoes causing the fruit to soften and mold. The Calgene scientists introduced an anti-sense DNA sequence into tomato seeds, whose plants produced tomatoes with decreased enzyme synthesis which prolonged shelf-life. Therefore, tomatoes can be left on the vine longer to ripen and still have sufficient time after harvesting to get them to market and remain edible without spoilage in the household. This is in contrast to the current practice of picking green tomatoes, shipping them, and then ethylene gassing them to turn red. Ethylene is a natural ripening agent in tomatoes.

In the Flavr-Savr tomato, a kanamycin-resistant (kan-r) marker gene was transferred with the anti-sense gene, and the FDA was concerned that the kan-r gene might produce an allergenic protein. The kan-r gene allows tomato cells to produce the protein APH (3') II (aminoglycoside-3' phosphotransferase) which inactivates the antibiotic, kanamycin. I have three tomato-allergic patients in whom I did parallel prick skin tests with fresh Flavr-Savr tomato, fresh commercial tomato, and a commercial tomato allergenic extract (Hollister-Stier, Spokane, WA). All 3 patients reacted equally with all three tomato products; there was no increase or decrease on allergenicity in the Flavr-Savr tomato.

Calgene did extensive pre-market safety testing to ensure that there was no increase in tomatine, the tomato toxin, that is present in similar form in all Solanaceae plants (tomato, potato, eggplant and peppers). Nutritional value also was the same as well.

Similar anti-sense polygalacturonase genes to prolong shelf life are being tried in tropical fruits - banana, papaya, and mangoes (1). Today, some 70% of the banana crop is lost through spoilage in transit to market. Therefore, a successful anti-spoiling banana would dramatically increase the marketable yield to either reduce the acreage and pesticide use needed or increase the supply and perhaps eventually reduce the unit cost and price of bananas.

Three other transgenic tomatoes are becoming available. 1) DNA Plant Technology has developed anti-sense ethylene transgenic tomatoes and red peppers which also retard ripening and prolong shelf-life. 2) Zeneca has introduced a higher glutenin content into its transgenic tomato which makes for a firmer tomato sauce. I have personally recently tasted all three of these transgenic tomato products and they are all very palatable with good flavor. However, they are still not quite like homegrown garden tomatoes. 3) Insect-resistant Bt tomato has been developed in Holland (1). *Bacillus thuringiensis* (Bt) produces a protoxin which when eaten by larval *Lepidoptera* in the midgut with high pH and proteases forms a potent insecticidal toxin. A gene segment of this Bt toxin has been introduced into tomato seeds and by Monsanto into cotton via recombinant DNA techniques. When the insect larva eats the tomato or cotton, the larva dies. It appears that mammalian gut lacks receptors for this insecticidal toxin, and mice and rats have shown no effect from being fed such transgenic Bt tomatoes or the Bt toxin itself. The Bt toxin is rapidly degraded in the mammalian gut.

Bt cottonseed produces an insect resistant cotton. Since cottonseed oil and fiber is used in animal feed, and cottonseed oil for direct human consumption, a safety concern for such foods in humans is being addressed. Bt protein in cotton plants is not glycosylated, so there is less risk of allergenicity.

One striking example of genetic transfer of allergenicity has recently been reported by Nordlee et al. (40). Soybean is a good widely used protein source in human and animal nutrition, but it is sulfur-poor, thus, not a complete nutrient. The 2S protein of Brazil nut contains 18% methionine making it a good candidate for sulfur supplementation in soybean. Therefore, a chimeric gene encoding 2S Brazil nut protein was transferred to soybean and the 2S protein was expressed in the transgenic soybean seeds. Because Brazil nut is a common allergen in tree-nut allergic persons, they tested sera from 8 tree-nut sensitive patients to see if they react to the Brazil nut-soy transgenic seeds.

RAST inhibition by extracts of Brazilnut and the transgenic seeds using a Brazil nut base showed almost identical inhibition slopes; this identified the Brazilnut gene expression in the transgenic seeds. Therefore, SDS-PAGE was used to separate proteins from Brazilnut, the 2S protein, transgenic soy seeds and normal soy seeds in 10-20% gradient gels, these were electroblotted on nitrocellulose and incubated with sera from 8 Brazilnut allergic patients. ^{125}I-labelled anti-human IgE was used to detect bound IgE. Purified 2S protein was recognized by 6/8 sera and by the same molecular weight region in Brazilnut and the transgenic seeds, but not by normal soybean seeds. This suggests that 2S protein which is the major allergen in Brazilnut is present in the transgenic soy seeds as a transferred allergen. It is, therefore, a likely possibility that a nut-sensitive person eating such a transgenic soy product would have an allergic reaction.

In contrast, Watanabe (41) has been successful in producing a hypo-allergenic rice by enzyme actinase treatment. This reduces the 16 kD rice allergen which almost completely disappeared, but the glutelin nutritional component of rice seed was unchanged. Anti-rice IgE antibodies measured by RAST scores in 6/8 children with eczema were completely removed. Clinically, eczema (skin area involved and severity scores) decreased with 6 weeks of hypoallergenic rice feeding, but rose again when normal rice was re-introduced.

Matsuda's group (42) has recently sequenced two related 14-16 kD amylase/trypsin inhibitor proteins in rice and with anti-sense genes reduced the expression of these proteins for a transgenic hypoallergenic rice. These will undergo safety and nutritive testing, and then clinical efficacy trials in Japan.

Conclusion

A great deal is known about the specific allergenic components of many foods which may be modified by genetic engineering in foods. However, the major consideration is cost in producing such allergen-free foods for the 1% of the food-allergic segment of the population. Labelling is the current FDA policy on potential allergens in transgenic foods, but this is expensive, as is the separation of transgenic foods from the general supply of such foods.

Biotech-engineered foods have tremendous potential in solving the problem of feeding the world's exponentially expanding population explosion; growing crops with greater yields in marginal tropical crop lands - drought and salt-resistant, and viral-, bacterial-, and insect-resistant foods. Finally, the exciting possibility is to introduce by means of recombinant DNA techniqes anti-disease vaccines into common foods, like bananas. Potential allergenicity is a concern that must be addressed in each new transgenic food, but it must not hinder the development of the tremendous potential for good from this biotechnology.

Literature Cited

1.   Kok, E.J.; Reynaerts, A.; Kuiper, H.A. *Trends Food Sci. & Tech.* **1993**, *41*, 42.
2.   Mitchison, N.A. *Eur. J. Immunol.* **1971**, *1*, 18.
3.   Mosmann, T.R.; Coffman, R.L. *Annu. Rev. Immunol.* **1989**, *7*, 145.
4.   Orgel, H.A.; Lenoir, M.A.; Bazaral, M. *J. Allergy Clin. Immunol.* **1974**, *53*, 213.
5.   Frick, O.L.; German, D.; Mills, J. *J Allergy Clin Immunol.* **1979**, *63*, 228.
6.   Baum, C.G.; Szabo, P.; Siskind, G.W. *J. Immunol.* **1990**, *145*, 779.
7.   Ishizaki, T.; Koizumi, R.; Ikemori, R. *Ann. Allergy*, **1987**, *58*, 265.
8.   Samuelsson, B. *Science*, **1983**, *220*, 568.

9.   Holgate, S.T.; Robinson, C.; Church, M.K. In *Allergy: Principles and Practice*, Editors, Middleton, E., Jr., Reed, C.E., Ellis, E.F. et al., 4th Ed., Mosby-Year Book, Inc., St. Louis, **1993**, p. 267.

10.   Lewis, T.; Grant, R.T. *Heart* **1924**, *11*, 209.

11.   Gleich, G.J.; Adolphson, C.R. *Adv. Immunol.* **1986**, *39*, 177.

12.   Frew, A.J.; Kay, A.B. *J. Allergy Clin. Immunol.* **1988**, *81*81, 1117.

13.   Dvoracek, J.E.; Yunginger, J.W.; Kern, E.B. et al. *J. Allergy Clin. Immunol.* **1984**, *73*, 363.

14.   Bonini, S.; Bonini, S.; Vecchione, A. et al. *J. Allergy Clin. Immunol.* **1988**, *82*, 462.

15.   Ermel, R.W.; Frick, O.L.; Reinhart, G.A. *J. Allergy Clin. Immunol.*, **1994**, *93*, 208.

16.   Sampson, H. A. In *Allergy: Principles and Practice*, Editors, Middleton, E., Jr.; Reed, C.E.; Ellis, E.F. et al., 4th Ed., Mosby-Year Book, Inc., St. Louis, **1993**, p. 1661.

17.   Chapman, M.D.; Arruda, L.K.; Sporik, R.B. et al. *FDA Conference on Scientific issues related to potential allergenicity in transgenic food groups*, Annapolis, MD, April 18-19, **1994**.

18.   Arruda, L.K.; Vailes, L.D.; Benjamin, D.C., Chapman, M.D. **1994**, Submitted.

19.   Berrens, L. *Int. Arch. Allergy* **1971**, *41*, 186.

20.   Bleumink, E.; Berrens, L.; Young, E. *Int. Arch. Allergy*, **1967**, *31*, 25

21.   Rountree, S.; Cogswell, J.J.; Platts-Mills; T.A.E., Mitchell, E.B. *Arch. Dis. Child* **1985**, *60*, 727.

22.   Johnsen, G.; Elsayed, S. *Mol. Immunol.* **1990**, *27*, 821.

23.   Kahlert, H.; Peterson, A.; Becker, W.M.; Schlaak, M. *Mol. Immunol.* **1992**, *29*, 1191.

24.   Barnett, D.; Howden, M.E.H. *Biochem. Biophys. Acta*, **1986**, *882*, 97.

25.   Burks, A.W.; Williams, L.W.; Connaughton, C. et al. *J. Allergy Clin. Immunol.* **1992**, *90*, 962.

26.   Dorion, B.J.; Burks, A.W.; Harbeck, R. et al. *J. Allergy Clin. Immunol.* **1994**, *93*, 93.

27.   Moroz, L.A.; Yang, W.H. *N. Engl. J. Med.* **1980**, *302*, 1126

28.   Herian, A.M.; Taylor, S.L.; Bush, R.K. *Int. Arch. Allergy* **1990**, *92*, 193.

29.   Ogawa, T.; Bando, N.; Tsuji, H. et al. *J. Nutr. Sci. Vitaminol.* **1991**, *37*, 555.

30.   Djurtoft, R.; Pedersen, H.S.; Aabin, B. et al., *Adv. Exp. Med. Biol.* **1991**, *289*, 281.

31.   Matsuda, T.; Nakamura, R. *Trends Food Sci. & Tech.* **1993**, *4*, 289.

32.   Sanchez-Monge, R.; Gomez, L.; Barber, D. et al. *Biochem. J.* **1992**, *281*, 401.

33.   Elsayed, S.; Apold, J. *Allergy* **1983**, *38*, 449.

34.   Frick, O. L.; Barker, S. *J. Allergy Clin. Immunol.* **1989**, *83*, 295.

35.   Bernheisel-Broadbent, J.; Strause, S.; Sampson, H.A. *J. Allergy Clin. Immunol.* **1992**, *90*, 622.

36.   Lehrer, S.B.; Ibanez, M.D.; McCants, M.L. et al. *J. Allergy Clin. Immunol.* **1990**, *85*, 1005.
37.   Valenta, R.; Breiteneder, H.; Pettenburger, K., et al. *J. Allergy Clin. Immunol.* **1991**, *87*, 677.
38.   Calkhoven, P.G.; Aalberse, M.; Koshte, V.L. et al. *Allergy* **1987**, *42*, 382.
39.   Valenta, R, Duchene, M., Pettenburger, K. et al. *Science* **1991**, *253*, 557.
40.   Nordlee, Taylor, S.L., *J. Allergy Clin. Immunol.* **1994**, *93*, 209.
41.   Watanabe, M. *Trends Food Sci. & Tech.* **1993**, *4*, 125.
42.   Matsuda, T.; Nomura, R.; Sugiyama, M.; Nakamura, R. *Agric. Biol. Chem.* **1991**, *55*, 509.

RECEIVED June 7, 1995

Chapter 10

Use of Plant Virus Genes To Produce Disease-Resistant Crops

Lynn E. Murry

Incyte Pharmaceuticals, 3330 Hillview Avenue, Palo Alto, CA 94304

Corn, tomatoes and cucurbits have been transformed successfully with viral coat protein (cp) genes to generate inheritable disease resistance. The effective amount of gene expression and specificity of the protection vary depending on virus classification and viral sequences used to engineer protection. Although mode of action is unknown, organs of transgenic plants contain less virus or viral protein than similar nontransgenic tissues exposed to natural infection. Engineered resistance is important because viruses can not be controlled chemically, and up to 20% of crop losses are attributed to viral diseases as single or mixed infections. A single corn virus may reduce yield 10-25%; however, when maize dwarf mosaic virus (MDMV) is combined with maize chlorotic mottle virus, mixed infection may cause 80-100% loss. Cp-mediated control of one of the viruses prevents such severe yield losses. Most agronomically important viruses are spread mechanically by farm equipment or biologically by insect or fungal vectors. Engineered cp expression provides protection against viral spread if it occurs early in the plant's life cycle. Use of biological or viral sequences, rather than chemicals, to control viruses or their vectors is safer for consumers and the environment.

Since Luther Burbank said, "We recently advanced our knowledge of genetics to a point where we can manipulate life in a way never intended by nature. We must proceed with the utmost caution in the application of this new found knowledge.", the safety of any change in a plant's genetics has been a major concern. The issue in 1906 was the safety of recombining plant genes using hybridization; today it is the safety of recombining genes through genetic engineering.

Even with hybridization, the available gene pool was quite limited, and crosses could only be made between closely related species. In the 1940s,

0097–6156/95/0605–0113$12.00/0
© 1995 American Chemical Society

polyploidization became feasible through application of chemical agents such as colchicine. Plant breeders attempted to use this "spindle anesthetic" (1) to double the chromosome number of infertile hybrids obtained through wide crosses. If cell cycle lengths were compatible and each chromosome had a suitable mate with which to pair, meiosis and fertility were restored. In the grasses, intergeneric crosses, although laborious, became possible; and sometimes, desired characteristics were moved into crop species where they were expressed.

With the advent of molecular biology, the number of species from which a "desirable" trait could be selected greatly expanded. In theory, any viral, prokaryotic or eukaryotic gene could be engineered to express in a recipient species. Thus, the question of safety is again under debate. It has triggered the entire range of cerebral responses, from carefully considered to entirely emotional, and various community and government groups have become embroiled in the discussions.

To date, the USDA has been most active in monitoring field tests of genetically engineered organisms; however, the FDA and EPA are becoming more involved as genetically engineered products approach commercialization. As is evident in Perils Against the Promise (2), scientists do not agree on which genes (or whether any genes) are safe. Moreover, ethical arguments on use of recombinant DNA in agriculture and medicine will undoubtedly proliferate well into the next century.

The focus of this paper is the safety of using genes from viruses to protect crop plants. It is surely one of the easier areas to address for three reasons. First, most animal or human pathogenic viruses are DNA viruses; most plant viruses are RNA viruses (3). When foods containing engineered nucleic acids and proteins are consumed, those molecules will be digested in the same way that the nucleic acids and proteins are digested in nonengineered foods. Second, foods engineered for viral resistance are likely to contain two to three orders of magnitude less virus RNA and protein than foods from infected plants (4). For example, a ripe tomato from the garden or the local grocery store which displays yellow, pin-point dots of tobacco or tomato mosaic virus (TMV) may contain as much as 2 μg coat protein (cp)/mg fresh weight. In contrast, a resistant tomato engineered with the TMV cp gene (5, 6) will contain only 2-5 ng cp/mg fresh weight. Protective levels of cp generally vary from ~ 0.01-0.2% of the total soluble plant protein (7). Third, several high-titer viruses may coexist within a single plant without recombining with one another or killing their host. This situation which is known as "mixed infection" and its consequences will be described in potatoes and corn below.

Fruits and vegetables from the home garden generally contain more virus than foods produced commercially. This is attributed to lower use of pesticides and insecticides, lack of knowledge of how viruses are spread, and abundant reservoirs of virus inoculum, usually in nearby weeds and ornamentals (8). In most cases, the fruits and vegetables from the garden or grocery are asymptomatic, and they are perfectly acceptable, even preferred, in taste and every other criterion. Considering the data from tomatoes presented above and the fact that viruses have always been present in our foods, consumption of engineered viral DNA, RNA and proteins should not be considered a food safety issue.

The engineering of virus resistance is based on earlier studies of the mechanism which produces "cross protection" viral strains. Cross protection involves challenging a crop with a mild strain of a virus to induce biochemical or molecular defense against a more virulent strain of the same virus. In a 1986 review, Fulton (9) stated that cross protected plants were resistant to inoculum 100 to 1000 fold greater than that required to infect a healthy plant. This practice was shown to be correlated with production of viral coat protein in the plant cell; however, cross protection is very labor intensive and is not practiced in modern agriculture.

Powell-Abel et al. (10) first demonstrated the concept of engineered coat protein protection in transgenic tobacco which expressed the cp of TMV. Since then, "first generation" technology (which may also include use of antisense or satellite sequences) has been demonstrated in various species with members of several different virus groups (11). The mechanism by which CP produces resistance has not been elucidated; however, several theories have been advanced. The leading theory is that excess coat protein prevents the virus from uncoating its genome, thereby preventing both translation and replication (11) in the plant cell. Although uncoating is actually the second event in the virus life cycle, it is the first logical point to attempt control of plant viruses since cell entry is usually aided by insects, fungi, humans and machinery.

Viral diseases are as hard to control in crop plants as they are in animals or humans. The four principle components which determine the severity of viral disease(s) are plant, virus, vector, and environment (Figure 1). The first component, the plant, may or may not have genetic resistance to a particular virus. Sometimes traits for virus resistance are present and can be used to prevent disease (12). More often these traits are multigenic, and the plant breeder has difficulty in moving them into new hybrids. If the virus has mutated and inherent genetic resistance no longer deters the new strain, then genetic engineering may allow the introduction of a single dominant trait to protect plants from the new strain and those older strains with closely related characteristics. The length of time that engineered resistance will be effective and whether it will hold up as well as plant genetic resistance is unknown.

Some viral diseases are contained using pesticides to control the vector component. The safety of these chemicals has been of concern for decades. The present generation of such chemicals (fungicides, herbicides, insecticides, and nematicides) is used in much lower concentrations and is generally more effective than those used earlier (13). The evolution of pesticide-resistant vector organisms, such as the whitefly which is spreading viral diseases all across the southern US (14), requires the agrochemical industry to produce and test new chemicals and formulations constantly. These chemicals are designed to control the vectors which damage crops directly by infesting, feeding on or laying their eggs on the plants and indirectly by spreading fungal or viral diseases.

The third component, environment, plays two opposing roles in the epidemiology of viral diseases. On the negative side, mild winters may allow infected plants to overwinter and serve as a source of inoculum for the next year's crop. The early season spread of viral diseases through young crop plants has the greatest potential to affect yield. Fig. 2 shows the projected yield of barley plants infected by barley yellow dwarf virus (BYDV) at different developmental stages

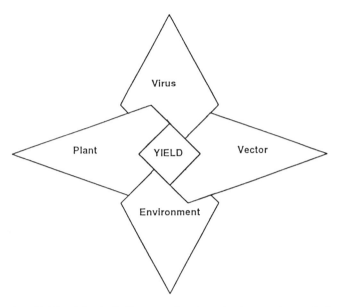

Figure 1.   Epidemiological diagram showing the interactions of the four components; plant, virus, vector, and environment which affect crop yield.

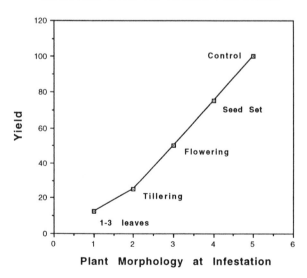

Figure 2.   Graph showing effect of barley yellow dwarf virus on yield of barley plants in various developmental stages. The younger the plant at infestation; the greater the yield loss.

over the course of the growing season. The earlier the disease manifests; the more severe the yield loss.

On the positive side, the build up of inoculum, spread and severity of viral diseases may be ameliorated by high temperatures, drought or other stresses during the growing season. This is true because the virus is absolutely dependent upon plant cellular machinery to reproduce. So, any stress which affects growth and metabolism of the plant cell induces natural defense responses (12), shuts down normal metabolism, and reduces viral replication. Secondarily, vectors may also suffer and show decline in numbers under stressful environmental conditions.

Estimates of crop losses due to the virus component vary tremendously depending on the crop (cf. 8). The cost directly attributed to the virus must be partitioned away from other diseases, pest and environmental damage. Estimates of corn virus losses in the United States are summarized in Fig. 3. These diseases do not reach epidemic proportions every year, but major losses in some portion of the growing region do occur with regularity (~every 3-5 years; 8). The most economically devastating situations are those ascribed to new strains or new mixed infections.

For example, maize dwarf mosaic virus (MDMV), the most common virus of corn, may exact a 10-25% yield loss. In addition, MDMV may be found in the mixed disease situation in the east, with maize chlorotic dwarf virus (MCDV) or in the west, with maize chlorotic mottle virus (MCMV). In corn lethal necrosis (CLN), which involves MDMV and MCMV, yield losses approach 90-100% (15). These viruses, which are vectored by different organisms and are members of different viral groups, seem to potentiate the disease state. Often, diagnosis of virus diseases is as difficult as their control. Fig. 4, shows some of the symptoms listed for the major corn viruses and points out the usefulness of antibody-based diagnostic kits. Now it is possible to determine which and how many viruses are contributing to visible symptoms.

Mixed viral infections are not uncommon; although they are probably underreported. Plants in the potato family seem to harbor multiple viruses quite well. One of the best examples is pepper where a single plant which may test positive for up to six viruses at a time. Generally new strains or new combinations of viruses seem to cause more extensive damage to crops than older, well-established disease relationships. However, it is also true that the diagnostic tools for assessing the extent and components of yield loss are now more widely available.

For example, in the first half of this century, agronomists began hybridizing potatoes to increase yield and to develop disease resistance. Their efforts and the use of pesticides and fungicides in commercial fields did increase yield; however, the real problem in potatoes was virus infection; and not just one virus, but usually a complex. Tubers, although asymptomatic were loaded with potato viruses X (PVX) and Y (PVY). The elucidation of two dominant genes for resistance to PVX have allowed breeding for virus resistance and seed certification programs to produce virus-free tubers and increase yield in Great Britain at least three-fold. There is excellent potential for PVX cp to confer similar protection and to increase or maintain high yields in other parts of the world. In Monsanto field trials, potatoes inoculated with PVX accumulated an average of 658 ng virus/mg fresh weight. By comparison, one resistant line of transgenic potatoes produced only 0.2

CORN VIRUS ECONOMICS

Loss	Virus	Vector
10-25%	MDMV	aphid
10-15%	MCMV	Diabrotica
90-100%	CLN	
10-25%	MCDV	leafhopper
50-80%	MDMV+MCD	
*	BYDV	aphid

* 50% of WI sweet corn plants infected in 1993

Figure 3. Chart showing effects of single or multiple viral diseases on corn yield.

SYMPTOMS OF CORN VIRUSES

	MDMV	MCMV	BYDV	MCDV
mosaic	x	x	x	
striping	x	x	x	x
chlorosis	x	x	x	x
dwarfing	x	x	x	x
reddening	x		x	x

Figure 4. Chart demonstrating the common symptoms attributable to virus diseases of corn and the need for diagnostic tests rather than visual assessment.

ng PVX cp/mg fresh weight. Only ~0.05% of the total protein in transgenic tubers was due to expression of the introduced gene (8, 9).

Vectors, particularly aphids, are very important in spreading potyviruses, the group to which both MDMV and PVY belong. Potyviruses are the largest group of plant viruses, and they infect many other important crop plants including tomatoes, squash, soybeans. Potyviruses are flexuous, rod-shaped viruses which consist of ~10,000 nucleotides of single-stranded RNA surrounded by ~200 units of a single, structural protein. The coat proteins in this family show 50-60% homology (18). Viral infection produces characteristic pinwheel-shaped structures in host cells and causes mosaic symptoms on the plant's leaves.

As previously illustrated with BYDV, potyviruses may exact a yield penalty whose severity corresponds to developmental stage at the time of infection. The Sandoz Seeds group chose to engineer resistance to MDMV using the cp gene from strain B because: 1) genetic resistance to MDMV in corn is multigenic and difficult to move into new hybrids, 2) strains of MDMV (whose coat proteins display about 88% homology) are found all the way from the Atlantic coast to the Rocky mountains, and 3) MDMV plays a pivotal role in the two mixed infections previously mentioned. The project (19) which produced MDMV-resistant/CLN tolerant transgenic plants took approximately three years. An abbreviated summary of that research follows.

The MDMV cp gene, which is located at the 3' end of the genome, was obtained as a cDNA clone from John Clark, Jr. at the University of Illinois. The cDNA was engineered into an agronomic cassette (or functional expression unit) consisting of a 35S cauliflower mosaic virus promoter, an alcohol dehydrogenase intron, the cp gene and a nopaline synthetase terminator. The cp cassette became part of a larger construction, named pZO1084, which contained a second, selection cassette carrying a promoter, the gene for neomycin phosphotransferase (npt, Fig. 5) and a terminator. The pZO1084 was introduced into sweet corn suspension cells by either particle bombardment or electroporation. Transformed cell lines were recovered via kanamycin selection for cells which could express NPT. These lines were propagated, tested for the presence and expression of the introduced genes, and regenerated into fertile plants.

The first transformed line, R-1084-1, expressed NPT at 90 ng/gram fresh weight and CP at 100 ng/mg fresh weight. This line, which gave rise to hundreds of genetically-identical plants, was tested in the growth chamber for its resistance to single and mixed viral inoculations. Although analysis of R-1084-1 tissues revealed that more than one copy of the cp gene is present, it appears that all copies integrated into a single site in the corn genome. This integration pattern simplifies tracking the gene, because all copies behave as a single, dominant marker when the transgenic is crossed into various commercial hybrids.

An experimental protocol for testing virus resistance was developed using sweet and field corn plants. This protocol used a known volume of inoculum per plant and resulted in the consistent appearance of disease symptoms in plants five to six days after inoculation. Fig. 6 shows the theoretical population distributions for plants displaying immunity, resistance, tolerance or susceptibility. The graphs from the MDMV viral challenges closely followed these patterns. It should be noted that sweet and field corn control populations were present in each experiment

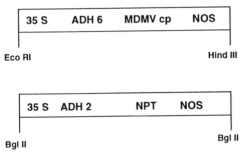

AGRONOMIC CASSETTE

SELECTABLE CASSETTE

Figure 5.   Agronomic and selectable marker cassettes used in the dual construct pZ01084 to transform sweet corn for resistance to MDMV.

VIRAL CHALLENGE PLOT

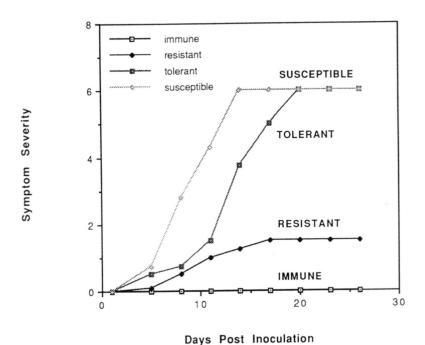

Days Post Inoculation

Figure 6.   Theoretical populations showing immunity, resistance, tolerance, and susceptibility to viral challenge over time.

and that all plants were at the same stage of development at the beginning of each experiment. The plants were challenged with two strains of MDMV, A and B, with MCMV, and with both viruses inoculated simultaneously. Plants were scored on a seven point scale where O was equivalent to no symptoms and 7, was lethally affected. The viral challenges were rated every few days, and the experiments continued until plant growth was limited by the dimensions of the growth chamber. Although growth chamber studies may be predictive, they cannot take the place of field trials (20) where the interactions of the transgene, pests and environment can be observed. Several generations of corn expressing cp have been tested in the field, and some of the populations which show superior resistance to the virus are being advanced.

The conclusions drawn from these growth chamber and field trial challenges were: 1) sweet and field corn controls were always susceptible to virus inoculation, 2) transformants were resistant both to MDMV-B and to the ten-fold more virulent MDMV-A, 3) transformants were susceptible to MCMV, and 4) transformants were quite tolerant to the mixed inoculation of MDMV-B + MCMV. The bottom line is that first generation technology which contains the cp of only one of the viruses participating in a mixed infection will limit yield loss.

Another public concern in commercializing virus resistant transgenic plants is the presence and expression of a selection cassette. The issue may be addressed by minimizing the effect of any particular selection cassette. First, the cassette can be designed 1) to produce less protein than an agronomic cassette, 2) to provide tissue specific expression, or 3) to be inducible (i.e., expressed) only when the selective agent is present and 4) to be removed via available biotechnological mechanisms (21). Second, selection cassettes may be based on genes which present little or no health concern. Although there are no scientific reasons on the grounds of human, veterinary or ecological safety to restrict the use of npt (22), there has been vocal concern in both the US and Europe over having an antibiotic resistance gene expressed in transgenic plant tissues. Surely the use of npt which was originally derived from the nonpathogenic, human gut bacterium, *Escherichia coli*, should be less of an issue than use of herbicide resistance. One of the great advantages of kanamycin selection is that the transgenic plants or their progeny need never see the active compound outside of the laboratory.

To date, the easiest solution to the selectable marker problem is to use cotransformation, to introduce separate constructs carrying agronomic and selectable cassettes. All cells which survive selection can be analyzed for presence of the agronomic construct and further work initiated. Progeny of transgenics which lose the selectable marker and in which only a single, integrated event is maintained and expressed can be advanced to field trials. Diagnostic tests such as PCR, ELISA, tissue prints, or dot blots can be used to follow segregation of the agronomic trait.

A slightly different situation, loss of the agronomic trait, has been documented in studies of transgenic tomatoes transformed with a dual construct containing TMV cp and npt genes. In that study, 18 F2 populations were advanced to field trials when their seeds were able to germinate on kanamycin-containing media. Following viral challenge, it was found that two of these populations scored as poorly as unprotected controls; they were fully susceptible to virus inoculation.

Reserve seed germinated readily on kanamycin-containing media, but diagnostic PCR revealed that both populations had lost the agronomic cp gene. While the loss of a selection cassette from a transgenic line may be considered a boon in some situations, the loss of an agronomic cassette (or its expression) from a commercial hybrid which has been advertised to carry a new resistance could be the subject of many, costly lawsuits. To prevent such occurrences, more basic research needs to be done on the directed integration and stable inheritance of heterologous genes.

Second generation antiviral technologies (defective interfering sequences; frameshift mutants; movement, nonstructural, or replicase genes) appear to be as effective and as safe as CP. The only problem with second generation technology is that it still takes one viral gene to protect against one virus. The great hope of both plant and animal biotechnologists is that third generation technologies will protect plants against a larger group of RNA, or DNA, viruses. Current studies which focus on using ribozymes to destroy a conserved region of a polymerase sequence or using a plantibody or inhibitor to inactivate reverse transcriptase or proteinase may actually accomplish this goal (cf. 23). These technologies, in their turn, will have to undergo sufficient testing to establish their efficacy and safety.

In summary, fruits and vegetables which carry and express virus genes are as safe to eat as any foods currently available. The use of plants which express resistance as a single dominant trait eliminates the need to accumulate resistance genes or try to move multigenic, often unlinked, traits from one hybrid or species to another. Control of viral diseases using engineered viral genes should allow 1) decreased use of the chemicals to control their vectors and 2) increased yield to the benefit of all consumers.

Acknowledgment. The research presented here was funded by Sandoz Seeds. I gratefully acknowledge the efforts of all my Sandoz colleagues, particularly those in the plant research, library, and patent groups, and thank them for numerous valuable discussions over the years.

Literature cited

1.   Darlington, C.D.; LaCour, L.F. *The Handling of Chromosomes*, Hafner Publishing Co. Inc., Darien, **1970**, pp. 91-92.
2.   Rissler, J.; Mellon, M. *Perils Amidst the Promise: Ecological Risks of Transgenics Crops in a Global Market* . Union of Concerned Scientists, Cambridge. **1993**, 92 pp.
3.   Wilson, T.M.A. *Bioessays,* **1989**, *70*, 179-186.
4.   Palukaitis, P. In *Risk Assessment in Genetic Engineering: Environmental Release of Organisms* . Levin, M.; Strauss, H., Eds.; McGraw-Hill, Inc. NY, **1991**, pp. 140-162.
5.   Nelson, R.S.; McCormack, S.M.; Delannay, X.; Dube, P.; Layton, J.; Anderson, E.J.; Kaniewska, M.; Proksch, R.H.; Horsch, R.B.; Rogers, S. G.; Fraley, R.T.; Beachy, R.N. *Bio/Technology,* **1988**, *6*, 403-409.
6.   Sanders, P.R.; Sammons, B.; Kaniewski, W.; Haley,L.; Layton, J.; LaVallee, B.J.; Delannay, X.; Tumer, N.E. *Molecular Plant Pathology*, **1992**, *82*, 683-690.

7. Grumet, R. *Hortscience*, **1990**, *25*, 508-513.
8. Matthews, R.E.F. *Plant Virology*, 3rd ed. Academic Press, NY, **1991**, 835 pp.
9. Fulton, R. W. *Ann. Rev. Phytopathol.*, **1986**, *24*, 67-81.
10. Powell Abel, P.; Nelson, R.S.; Barun, D.; Hoffmann, N.; Rogers, S.G.; Fraley, R.T.; Beachy, R.N. *Science*, **1986**, *232*, 738-743.
11. Beachy, R.N. In *Biotechnology in Plant Disease Control*, Wiley-Liss, Inc. **1993**, pp. 89-104.
12. Fraser, R.S.S. *CRC Crit. Rev. Plant Sci.*, **1986**, *3*, 257-294.
13. Riviere, J.E. *Newsweek*, 8 August **1994**, p.8.
14. Brown, J.K.; Bird, J. *Plant Disease*, **1992**, *76*, 220-225.
15. Niblett, C.L.; Claflin, L.E. *Plant Dis. Rep.*, **1978**, *62*, 15-19.
16. Lawson, C.; Kaniewski, W.; Haley, L.; Rozman, R.; Newell, C.; Sanders, P.; Tumer, N.E. *Bio/Technology*, **1990**, *8*, 127-134.
17. Kaniewski,W.; Lawson, C.; Sammons, B.; Haley, L.; Hart, J.; Delanney, X.; Tumer, N.E. *Bio/Technology*, **1990**, *8*, 750-754.
18. Reichmann, J.L.; Lain, S.; Garcia, J.A. *J. Gen. Virol.*, **1992**, *73*, 1-16.
19. Murry L.E.; Elliott, L.G.; Capitant, S.A.; West, J.A.; Hanson, K.K.; Scarafia, L.; Johnston, S.; DeLuca-Flaherty, C.; Nichols, S.; Cunanan, D.; Dietrich, P.S.; Mettler, I.J.; Dewald, S.; Warnick, D.; Rhodes, C.; Sinibaldi, R.M.; Brunke, K.J. *Bio/Technology*, **1993**, *11*, 1559-1564.
20. Dale, P.J.; Irwin, J.A.; Scheffler, J.A. *Plant Breeding*, **1993**, *111*, 1-22.
21. Yoder, J.l.; Goldsbrough, A.P. *Bio/Technology*, **1994**, *12*, 263-267.
22. Flavell, R.B.; Dart, E.; Fuchs, R.L.; Fraley, R.T. *Bio/Technology*, **1992**, *20*, 141-144.
23. Wilson, T.M.A. *Proc. Nat. Acad. Sci. USA*, **1993**, *90*, 3134-3141.

RECEIVED June 7, 1995

Chapter 11

Molecular Cloning of Cereal Cystatins and Evaluation of Their Antiviral and Antipest Effects

S. Arai, M. Kuroda, I. Matsumoto, H. Watanabe, and K. Abe

Department of Agricultural Chemistry, University of Tokyo, Bunkyo-ku, Tokyo 113, Japan

Cystatins refer to proteinaceous cysteine proteinase inhibitors. The oryzacystatin we found in rice is known to be the first well-defined cystatin of plant origin. Recently, we elucidated similar cystatins from corn and wheat by molecular cloning and expression studies. These small proteins contain approximately 100 amino acid residues. Each has a central Gln-Val-Val-Ala-Gly or related region which may be involved in binding with cysteine proteinases such as papain. Wheat cystatin, as well as oryzacystatin, is also characterized by having an inhibitory effect on human viruses, e.g., poliovirus, and crop insects, e.g., adzuki bean weevil which has a cysteine proteinase in its digestive tract. Our finding of such defensive functions of cereal cystatins prompted us to regenerate transgenic rice in which cystatin genes had been introduced. The expression of the introduced genes has already been confirmed. The possible utility of these new rice cultivars will be discussed.

Cystatins are proteinaceous cysteine proteinase inhibitors belonging to the cystatin superfamily (1). We found a cystatin in the seed of rice, *Oryza sativa* L., and have named it oryzacystatin (OC) after detailed studies on gene and protein levels (2,3). Subsequently, extensive studies were made to characterize the gene and protein structures of OC (5-8). OC has thus become the first well-defined phytocystatin (9).

This first phytocystatin was renamed oryzacystatin-I (OC-I). Since then, we found a second oryzacystatin, termed OC-II, in rice seeds by cloning with OC-I cDNA as a probe for hybridization (10). Both OC-I and OC-II have their target enzymes, oryzain α, β and γ, which we have cloned using aleurain cDNA as a probe (11) and characterized as gibberellin-inducible cysteine proteinases endogenously occurring in rice seeds (12-14). Phytophysiologically, OC-I and OC-II are involved in regulating germination which is accelerated by oryzain α, β and γ in the presence of gibberellins (14,15).

Besides the regulatory function, some protective effects may exist in OC. One example is its antiviral effect, since egg cystatin, a well-defined animal cytatin, has an inhibitory effect on the proliferation of poliovirus (16), a representative of picorna

0097–6156/95/0605–0124$12.00/0

viruses, which is characterized by having a cysteine proteinase, 3C proteinase, as its essential factor (17). We also found that OC-I, OC-II and a truncated mutant of OC-I (5) elicit an antiviral function as will be discussed later. Another protective effect is observed when crop insect pests are killed by feeding on diets with added OC. This is probably due to the fact that OC inhibits the growth of insects which generally have cysteine proteinases as digestive enzymes in their guts.

To further elucidate the antiviral and antipest effects of cereal cystatins in general, we have attempted to determine cystatins of corn (19), wheat and soybean origins primarily by molecular cloning. In the present paper, however, the procedures for the cloning are described mainly for OC-I. For antiviral and antipest effects of cereal cystatins, our discussions also focus on those of OC. Finally, we would like to discuss in some measure the possibility of creating new antipest crops, with special emphasis on construction of transgenic rice with enhanced cystatin expression.

Molecular Cloning for Characterization

Rice, *Oryza sativa* L., is a staple food resource in Japan as well as in many countries of the world. A great deal of effort has long been made world-wide to develop new rice cultivars of nutritional importance. Special attention has been paid to improving the quantity and quality of storage proteins, *e.g.*, glutelin, in the seed. However, little is known about molecular events that may take place for processing of preproglutelin into mature protein during ripening of seeds and for its degradation during germination. A working hypothesis is that either one or both of these two proteolytic events are controlled by, among others, cysteine proteinases and their natural inhibitors such as cystatins.

The seeds of all japonica-type rice cultivars investigated contain OC-I at 2-3 mg/kg when the content was assayed using an antibody raised against a purified OC-I preparation (7). OC-I purified from a typical rice cultivar, Nakate-shinsenbon, was a 11.5 kDa protein with pI 5.3. It did not inhibit serine and aspartic proteinases but efficiently inhibited the representative cysteine proteinases, papain and ficin (12,13). A cysteine proteinase in germinating rice seeds, which is referred to as oryzain α (14), can be almost stoichiometrically inhibited by OC-I (13).

In our cloning experiment, a cDNA clone for OC-I was isolated from a λgt10 cDNA library of immature rice seeds (cultivar Nipponbare) by screening with two synthesized oligonucleotide probes based on the OC-I partial amino acid sequences Lys-Pro-Trp-Met-Asp-Phe and Lys-Pro-Val-Asp-Ala-Ser which had been chemically determined. A nearly full-length cDNA clone was obtained encoding 102 amino acid residues. The amino acid sequence of OC-I deduced from the nucleotide sequence of this cDNA clone was significantly homologous to those of animal cystatins (4). Interestingly, OC-I contained the pentapeptide sequence Gln-Val-Val-Ala-Gly conserved as a target enzyme-binding site in most animal cystatins belonging to the cystatin superfamily (Figure 1).

We also isolated a genomic DNA clone encoding OC-I from a λEMBL3 phage library constructed with *Sau*3A I partial digests of rice seed chromosomal DNA ; the OC-I cDNA was used as a probe for the isolation (6). The restriction map of the isolated DNA fragment was consistent with the pattern of the genomic Southern blot analysis using the cDNA probe and, consequently, the DNA fragment is considered to be a single-copy gene. The OC-I gene was 1.4 kb long and composed of three exons and two introns. The first intron spanning 336 bp intervened between Ala-38 and Asn-39 (Figure 2). The second intron, 372 bp, existed in the 3'-noncoding region at the G nucleotide residue next to the stop codon TAA. S1 nuclease mapping showed the major transcription start point (tsp) at the nucleotide residue A, 104 bp upstream from the start codon (ATG). Typical CAAT and TATA box sequences were found in the 5'-upstream region of the tsp. The nucleotide sequences around the TATA box,

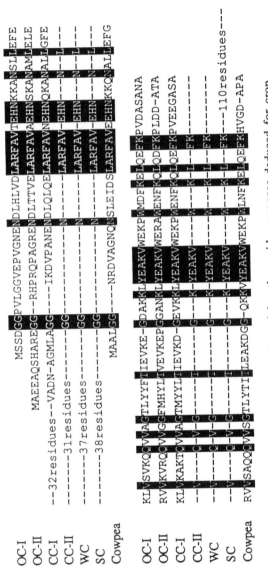

Figure 1. Complete and partial amino acid sequences deduced for crop cystatins (4, 10, 19, 44).

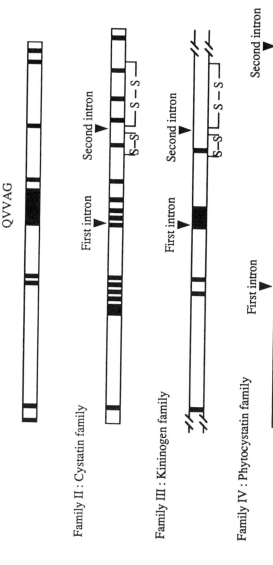

Figure 2. Illustration showing homology among cystatins belonging to the three established families and a new family proposed for phytocystatins (9). Positions of commonly conserved amino acid residues are indicated in black. Intron positions are shown by wedge-shaped marks.

Table I. Anti-polioviral activity of various inhibitors (24)

Inhibitors	Polio virus (virus yield, PFU/ml)
Oryzacystatin I	1.0×10^4
Oryzacystatin II	1.0×10^5
Truncated oryzacystatin I	1.2×10^4
Egg white cystatin	1.0×10^4
Soy bean trypsin inhibitor	4.1×10^7
E-64	5.4×10^7
E-64C	3.3×10^7
Loxistatin	3.5×10^7
Pepstatin	3.4×10^7
None (control)	3.5×10^7

Table II. Anti-polioviral activities of cystatins used at various concentrations (24)

Cystatins	Concentrations (nmol/ml)	Virus yield (PFU/ml)
Oryzacystatin-I	8	1.0×10^4
	4	1.2×10^6
	2	1.0×10^7
	1	3.0×10^7
Oryzacystatin-II	8	1.0×10^5
	4	5.0×10^6
Truncated oryzacystatin-I	2	6.5×10^6
	1	2.3×10^7
	2	1.2×10^5
	1	9.0×10^6
	0.5	1.6×10^6
	0.25	2.8×10^7
Egg white cystatin	8	1.0×10^4
E-64	8	5.4×10^7
None (control)	-	3.5×10^7

the tsp, the start codon, and the stop codon essentially matched the consensus sequences of other higher plant genes. A most characteristic point was that the intron boundaries of the OC-I gene were quite different from those of animal cystatin genes.
We isolated cDNA clones for OC-II by screening with the OC-I cDNA probe (10). The newly isolated cDNA clone encoded 107 amino acid residues whose sequence was 55% homologous to that of OC-I. OC-II, as well as OC-I, had no disulfide bonds and, therefore, could be classified as a family-1 cystatin member or a stefin ; however, the amino acid sequences of OC-I and OC-II resembled those of family-2 rather than family-1 cystatin members and seemed to be chimeric. We have also succeeded in cloning cystatins from corn (19), wheat, and soybean (Figure 1). Any of OC-I, OC-II and other cereal cystatins seemed to be a chimera of family-1 and family-2 cystatins, and it is proposed that these phytocystatins be classified into family-4 as a new category of the cystatin superfamily (9) (Figure 2).

Antiviral Effect

There are a variety of human viruses including picornaviruses which enter the intestine *per os*. Picornavirus usually has a positive-stranded genomic RNA and, when entering animal cells, synthesizes a high-molecular-weight polyprotein, which is subsequently divided into a number of viral components (17,20-22). A characteristic point in this molecular event is that the cleavage of polyprotein is carried out by virus-encoded proteinases, one of which, known as 3C protease, has a cysteine residue in its catalytic center (23). It follows that inhibition of this viral cysteine proteinase by cystatin would lead to stopping the proliferation of virus particle in the infected cells.
   In order to verify this, we selected poliovirus as the model target. Poliovirus, which belongs to the picornavirus, causes a severe disease and, therefore, has long been studied in detail. Cultured Vero cells (originating from monkey kidney) infected with poliovirus were incubated in the presence or absence of OC-I, OC-II, and N-terminally truncated OC-I (5), and the resultant virus yield was represented in terms of plaque-forming units (PFU). The result was that the PFU value obtained in the presence of each of the cystatins remained much lower (approximately 1/100) than that obtained in its absence (24) (Table I), indicating that these proteins have an antiviral effect to a greater or lesser extent depending on the concentration. Recently, we have found that wheat cystatin has a similar antiviral effect (unpublished data). It should be noted that E-64, a low-molecular-weight cysteine proteinase inhibitor of microbial origin (25), is not effective (24) (Table II). This result suggests that the use of a cystatin as a proteinaceous substance is important in obtaining an antiviral effect. Though nothing is known experimentally concerning the mechanism involved, it is speculated that cystatin molecules enter the infected cells by internalization to inhibit the molecular event in which the polyprotein is processed into functional products (16, 24).
   The potyvirus group, which is well known as a major crop pest, has a similar cysteine proteinase-dependent system for proliferation (26). Garcia *et al.* has reported that, in an experiment using the *in vitro* translation system, cystatin can block the processing of the potyvirus polyprotein, but leupeptin can not (27). These findings also suggest the significance of using such proteinaceous inhibitors. It is thus important to take notice of phytocystatins as useful factors for molecular breeding of virus resistant crop cultivars, as well as of functional food materials with an antiviral effect for human consumption (28).

Antipest Effect

The crop damage inflicted by invading insect pests is a serious problem for agricultural and food industries. A great deal of effort has been directed toward establishing a countermeasure against the pests. Indeed, a number of low-molecular-weight chemicals as pesticides have been developed to occupy a central position in agriculture. Often insects develop resistance to these pesticides when they are used over a long period of time. Thus new chemicals are developed one after another in a ridiculous circle. Residual chemicals in foodstuffs may also cause a serious safety problem.

Insects as well as many other invaders of crops devour them. They utilize proteins stored by crops by digesting them with their intralumen proteases. This means that blocking the digestive proteases of insects could cause protein starvation, often with lethal results. Therefore, much attention has been paid to the use of proteinase inhibitors for this blocking. This strategy arises from the fact that plant seeds *per se* contain a variety of proteinase inhibitors probably as bioprotectants (29). A number of naturally occurring proteinase inhibitors have been isolated, characterized and checked for effectiveness; most of them are serine proteinase inhibitors including trypsin inhibitors (29).

It has been reported that serine proteinase inhibitors are effective in regulating the protease activity of a crude extract from the gut of insects such as Lepidoptera (30). This finding is supported by several of the following studies confirming that feeding insects on a diet with added plant serine protease inhibitor can retard their growth to a significant extent (31-33). Most crops are still damaged by invading insect pests. This suggests that the target should be set preferably at cystatins which are safe for human consumption and yet block cysteine proteinases as they act in the gut of insect pests.

The occurrence of gut cysteine proteinases has been deduced by studies with hematophagous Hemiptera (30). Recently, it was found that a Coleoptera insect inflicting great damage to beans has a cysteine proteinase in the digestive tract (34-36). This finding has now become a testimony to the significant involvement of this type of proteinase. Since, however, there is no phytocystatin available so far, the only way has been to use E-64 as a possible gut cysteine proteinase inhibitor. According to published data from experiments with common bean beetle (37), Colorado potato beetle (38), and cowpea weevil (39), this inhibitor, added to some artificial diets, is significantly effective in retarding their growth. The data also suggest the possible effectiveness of phytocystatins of cereal origin.

We have completed a system for overexpression of OC by introduction of its gene into *Escherichia coli* (5) making it easy to use recombinantly produced OC for the antipest purpose. For this, we selected *Callosobruchus chinensis* as an invader of azuki and cowpea seeds, and *Riptortus clavatus* as a representative soybean pest. Feeding each of these insects on their respective diets with added OC-I or OC-II, we found that the inhibition clearly showed a growth-retarding effect, even with a lethal result when fed at higher concentrations (unpublished data). We also tested the inhibitory effect on the growth of *Sitophilus zeamais*, a notorious invader of rice seeds. The result was almost similar in that both OCs can retard the growth of this insect and even kill it at the larval stage. The results also indicate that rice *per se* has such an inhibitor as a self-protectant, though its quantity may be insufficient for complete protection from an attack. Recently, Chen *et al.* reported that recombinant OC-I showed a similar growth-inhibitory effect on *Tribolium castaneum* (40) and that the rice extract can block its gut protease (41). It is probably a generally observable phenomenon that different phytocystatins inhibit different cysteine proteinases of insect gut origin.

There is, thus, an experimental background warranting the effectiveness of phytocystatins as potential antipest substances. The effect, however, depends on the quantity or concentration of each cystatin used, since different insects require different lethal doses. For example, the effective concentration of OC-I is 10% in the diet for *Tribolium castaneum* (40) which is almost two-orders higher than that for *Callosobruchus chinensis* (unpublished data). Such a difference may be due to a difference in specificities of cystatins toward a insect gut cysteine proteinase, also reflecting a difference in terms of how much the cysteine proteinase is involved in the digestive process. It suggests the significance of studying cysteine proteinases of insects themselves as well as cystatins. Despite that, the study has been conducted only by using crude extracts from insect guts; no detailed data have been presented from experiments using purified cystatins and no attempts have been made to explain this from the standpoint of enzymology and molecular cloning.

We then started cloning a cysteine proteinase from *Drosophila melanogaster* as a model insect material and obtained several positive genomic DNA clones. The protein deduced to be encoded by one of those clones was highly homologous to cathepsin L in terms of amino acid sequence (unpublished data). *In situ* hybridization to look at stage- and tissue-specific expression showed that the mRNA encoding this protein is strongly expressed in the digestive organ of *D. melanogaster* at its embryonic and larval stages. The result may probably be the first that proves the existence of at least one cysteine proteinase in the gut on a molecular level. A similar experiment with *S. zeamais* is under way to confirm the involvement of a cysteine proteinase in its digestive process and to verify the inhibitory effect of added phytocystatins on this digestive enzyme.

Attempts to Construct a Novel Trangenic Rice Cultivar

It is probable that, unlike trypsin inhibitors, cystatins have no anti-nutritional effect on the human digestive system which does not contain cysteine proteinases. The construction of transgenic rice with enhanced cystatin expression would thus appear to be safe from a human standpoint.

In order to establish transgenic systems for development of crops with novel protective functions, we attempted to construct a chimeric gene encoding an oryzacystatin (OC)-β-glucuronidase (GUS) fusion protein with two activities. GUS reporter enzyme, which has been well-defined, is easily detected for its activity and, therefore, its expression can be used as a selection marker (42). In order to construct transgenic rice plant with an OC-GUS fusion gene, we first introduced it into rice protoplasts by electroporation, together with a marker gene conferring hygromycin-resistance (pUC-HPH). In a transient assay using the transfected protoplasts, both OC and GUS activities were detected. The GUS activity was higher when the OC-GUS fusion protein was expressed than when GUS was expressed alone. Next, in order to isolate stable transformants, hygromycin-resistant calli were selected. Forty one out of 116 hygromycin-resistant calli expressed a 2.2 kb mRNA transcribed from the chimeric gene and their extracts exhibited the activities of both OC and GUS. Finally, the transgenic calli were regenerated into rice plants whose tissues (leaves, roots, and seeds) exhibited GUS activity derived from the fusion protein (Hosoyama, H. Plant Cell Reports, *in press.*).

On the other hand, since corn cystatin (CC) shows a wide inhibitory spectrum against various cysteine proteinases (43), we established transgenic rice plants by introducing CC cDNA under CaMV35S promoter to obtain an antipest rice plant. This was based on the observations that many insect pests have cysteine proteinases probably as their digestive enzymes and also that OC, as an intrinsic rice cystatin, shows a narrow inhibition spectrum (43) and its amount in ordinary rice seeds is insufficient to inhibit all the cysteine proteinases of invading pests (7). The transgenic rice plants we established contained a high level of CC mRNA and CC protein in both seeds and leaves, and the CC protein content of each seed reached *ca.* 2% of its heat-soluble protein (unpublished data). We also recovered CC activity from the seeds and found that the CC fraction efficiently inhibited both papain and cathepsin H, whereas the corresponding fraction from ordinary, non-transformed rice seeds showed a much lower or undetectable inhibitory activity against these cysteine proteinases. Furthermore, CC prepared from the transgenic rice plant showed protein inhibitory activity against proteases occurring in the gut of the insect pest, *Sitophilus zeamais.* These results present a novel strategy for molecular breeding of safe crop plants with antipest activity.

It is added that the safety of cereal cystatins may probably warranted by our long experience of eating lots of rice, wheat, corn, and soybean day by day, all of which contain cystatins. However, we need further experimentation to confirm in a final manner the safety of the transgenic rice whose cystatin content has been enriched artificially.

References

1. Barrett, A. J., Rawlings, N. D., Davies, M. E., Machleidt, W., Salvesen, G., Turk, V.: In *Proteinase Inhibitors*; Barrett, A. J., Salvesen, G., Eds.; Elsevier Scientific Publishing Co.: Amsterdam, **1986**, Vol. 12; pp 515-569.
2. Abe, K.; Arai, S. *Agric. Biol. Chem.* **1985**, *49*, 3349-3350.
3. Abe, K.; Kondo, H., Arai, S. *Agric. Biol. Chem.* **1987**, *51*, 2763-2768.
4. Abe, K.; Emori, Y., Kondo, H., Suzuki, K., Arai, S. *J. Biol. Chem.* **1987**, *262*, 16793-16797.
5. Abe, K.; Emori, Y., Kondo, H., Arai, S., Suzuki, K. *J. Biol. Chem.* **1988**, *263*. 7655-7659.
6. Kondo, H.; Emori, Y., Abe, K., Suzuki, K., Arai, S *Gene* **1989**, *81*, 259-265.
7. Kondo, H.; Abe, K., Arai, S. *Agric. Biol. Chem.* **1989**, *53*, 2949-2954.
8. Kondo, H.; Abe, K., Fujisawa, I., Tada, T., Arai, S. *Agric. Biol. Chem.* **1990**, *54*, 2153-2155.
9. Abe, K.; Kondo, H., Watanabe, H., Emori, Y., Arai, S. *Biomed. Biochim. Acta* **1991**, *50*, 637-641.
10. Kondo, H.; Abe, K., Nishimura, I., Watanabe, H., Emori, Y., Arai, S. *J. Biol. Chem.* **1990**, *265*, 15832-15837.
11. Rogers, D. J.; Dean, D., Heck, G. R. *Proc. Natl. Acad. Sci. USA* **1985**, *82*, 6512-6518.
12. Abe, K.; Kondo, H., Arai, S. *Agric. Biol. Chem.* **1987**, *51*, 1509-1514.
13. Arai, S.; Hosoyama, H., Abe, K. *Agric. Biol. Chem.* **1988**, *52*, 2957-2959.
14. Watanabe, H.; Abe, K., Emori, Y., Hosoyama, H., Arai, S. *J. Biol. Chem.* **1991**, *266*, 16897-16902.
15. Watanabe, H.; Abe, K., Arai, S. *Biosci. Biotech. Biochem.* **1992**, *56*, 1154-1155.
16. Korant, B. D.; Brzin, J., Turk, V. *Biochem. Biophys. Res. Comm.* **1985**, *127*, 1072-1076.
17. Lawson, M. A.; Semler, B. L. *Curr. Top. Microbiol. Immunol.* **1990**, *161*, 49-80.
18. Arai, S.; Watanabe, H., Emori, Y., Abe, K. *J. Biochem.* , **1991**, *109*, 294-298.
19. Abe, M.; Abe, K., Kuroda, M., Arai, S. *Eur. J. Biochem.* **1992**, *209*, 933-937.
20. Palmenburg, A. C.; Rueckert, R. R. *J.Virol.* **1982**, *41*, 244-249.
21. Klump, W.; Marquardt, O., Hofschneider, P. H. *Proc. Natl. Acad. Sci. USA* **1984**, *81*, 3351-3355.
22. Hanecak, R.; Semler, B. L., Ariga, H., Anderson, C. W., Wimmer, E. *Cell* **1984**, *37*, 1063-1073.
23. Ivanhoff, L. A.; Towatari, T., Ray, J., Korant, B. D., Petteway, S. R. *Proc. Natl. Acad. Sci. USA* **1986**, *83*, 5392-5396.
24. Kondo, H.; Ijiri, S., Abe, K., Maeda, H., Arai, S. *FEBS Lett.* **1992**, *299*, 48-50.
25. Hanada, K.; Tamai, M., Tamaguchi, M., Ohmura, S., Sawada, J., Tanaka, I. *Agric. Biol. Chem.* **1978**, *42*, 523-526.
26. Riechman, J. L.; Lain, S., Garcia, J. A. *J. Gen. Virol.* **1992**, *73*, 1-16.
27. Garcia, J. A.; Cervera, M.T., Riechman, J. L., Lopez-Otin, C. *Plant Mol. Biol.* **1993**, *22*, 697-701.

28. Swinbanks, D.; O'Brien, J.*Nature* **1993**, *364*, 182.
29. Ryan, C. *Annu. Rev. Phytopathol.* **1990**, *28*, 425-449.
30. Applebaum, S.W.: Biochemistry of Digestion; In *Comprehensive Insect Physiology, Biochemistry and Pharmacology*; Kerkut, G. A., Gilbelt, L. I. Eds.; Pergamon Press: New York, **1985**, Vol. 4; pp279-311.
31. Hilder, V. A.; Gatehouse, A. M. R., Sheerman, S. E., Barker, R. F., Boulter, D. *Nature* **1987**, *330*, 160-163.
32. Johnson, R.; Narvaez, J., An, G., Ryan, C. *Proc. Natl. Acad. Sci. USA* **1989**, *86*, 9871-9875.
33. Oppert, B.; Morgan, T. D., Culbertson, C., Kramer, K. *J. Comp. Biochem. Physiol.* **1993**, *105C*, 379-385.
34. Kitch, L. W.; Murdock, L. L. *Arch. Insect Biochem. Physiol.* **1986**, *3*, 561-575.
35. Murdock, L. L.; Brookhart, G., Dunn, P. E., Foard, D. E., Kelley, S.,Kitch, L., Shade, R. E., Shukle, R. H., Wolfson, J. L. *Comp. Biochem. Physiol.* **1987**, *87B*, 783-787.
36. Wieman, K. F.; Nielsen, S. S. *Comp. Biochem. Physiol.* **1988**, *89B*, 419-426.
37. Hines, M. E.; Nielsen, S. S., Shade, R. E., Pomeroy, M. A. *Entomol. Exp. Appl.* **1990**, *57*, 201-207.
38. Wolfson, J. L.; Murdock, L. L. *Entomol. Exp. Appl.* **1987**, *44*, 235-240.
39. Murdock, L. L.; Shade, R. E., Pomeroy, M. A. *Environ. Entomol.* **1988**, *17*, 467-469.
40. Chen, M. S.; Johnson, B., Wen, L., Muthukrishnan, S., Kramer, K. J., Morgan, T. D., Reeck, G. R. *Protein Express. Purification* **1992**, *3*, 41-49.
41. Liang, C.; Brookhart, G., Feng, G. H., Reeck, G. R., Kramer, K. J. *FEBS Lett.* **1991**, *278*, 139-142.
42. Jefferson, R. A.; Kavanagh, T. A., Bevan, M. W. *EMBO J.* **1987**, *6*, 3901-3907.
43. Abe, M.; Abe, K., Iwabuchi, K, Domoto, C., and Arai, S. *J. Biochem.* **1994**, *116*, 488-492.
44. Fernandes, K. S. V.; Sabelli, P. A., Barratt, D. H. P., Richardson, M., Xavier-Filho, J., and Shewry, P. R. *Plant Mol. Biol.* **1993**, *23*, 215-219.

RECEIVED June 7, 1995

Chapter 12

Safety Assessment of the *Bacillus thuringiensis* Insecticidal Crystal Protein CRYIA(b) Expressed in Transgenic Tomatoes

H. P. J. M. Noteborn[1], M. E. Bienenmann-Ploum[1],
J. H. J. van den Berg[2], G. M. Alink[2], L. Zolla[3], A. Reynaerts[4],
M. Pensa[5], and H. A. Kuiper[1]

[1]Department of Risk Assessment and Toxicology, RIKILT-DLO,
P.O. Box 230, 6700 AE Wageningen, Netherlands
[2]Department of Toxicology, Agricultural University Wageningen,
6700 AE Wageningen, Netherlands
[3]Department of Environmental Sciences, University La Tuscia,
Viterbo, Italy
[4]Plant Genetic Systems, Jozef Plateaustreet 22, Ghent, Belgium
[5]SME Ricerche SCPA, La Fagianeria, Piana di Monte Verna, Italy

Transgenic Bt-tomatoes were produced which contain a gene of *Bacillus thuringiensis* *(Bt)* encoding the CRYIA(b) protein, and a selectable marker gene expressing neomycin phosphotransferase II (NPTII). *In vivo* and *in vitro* experiments in gastro-intestinal tissues from rodents, rhesus monkey, and humans indicated the absence of specific binding sites for CRYIA(b) protein. Short-term toxicity testing with CRYIA(b) protein revealed no adverse effects in laboratory animals, and no evidence was found for immunotoxicity of the protein. Chemical analysis did not show major changes in nutritional composition of transgenic Bt-tomatoes compared to that of control lines. Moreover, levels of the glycoalkaloid α-tomatine were similar. Preliminary analysis of the results of a 91 day feeding trial of transgenic Bt-tomatoes in rats did not reveal any signs of adverse effects.

The development of genetic engineering techniques has resulted in important applications in the field of plant breeding. Traits for insect and virus resistance, cold/drought resistance, herbicide tolerance and delayed ripening of fruits have been introduced in a number of crop plants *(1-6)*. Evaluation of transgenic food crops has for a great part been limited to yield and open field behavior with particular attention to deliberate release of genetically modified organisms in the environment. Food safety testing of modified crops is still in an explorative phase. Recently, the FLAVR SAVR™ tomato (Calgene) with delayed ripening properties was approved for market introduction in the USA, and many new products have reached the phase of market introduction. Testing strategies for the establishment of the safety of genetically modified foods for the consumer have been developed by various national governments and international organizations (7). Evaluation of first experimental experiences in this field is of great importance for validation of proposed guidelines.

In 1991 an EU co-sponsored research project was initiated in the framework of the FLAIR (Food-Linked Agro-Industrial Research) program entitled '*Opportunities of Transgenic Food Crops for the Consumer and the Food Industry in the Community*' (contract no. AGRF-0039). The project concerns the molecular/biochemical and toxicological characterization of tomato transformants and derived fruit, which were genetically modified by the introduction of a gene encoding an insecticidal crystal protein, CRYIA(b), from *Bacillus thuringiensis* (for the classification of the *B. thuringiensis* toxins see (8)). Sprays of spore-crystal mixtures of the bacterium have been used during the last 50 years as a biological insecticide (9). An easily detectable marker gene, encoding neomycin phosphotransferase (NPTII) was co-introduced into the plant genome for selection purposes.

Partners in the project are Plant Genetic Systems N.V., Ghent, Belgium (coordinator); the State Institute for Quality Control of Agricultural Products (RIKILT-DLO), Wageningen, the Netherlands, in cooperation with the Department of Toxicology of the Agricultural University of Wageningen, and with the Department of Environmental Sciences of the University La Tuscia, Viterbo, Italy; SME Ricerche SCPA, La Fagianeria, Piana di Monte Verna, Italy, and the University of Genova, DIBE, Genova, Italy. This paper describes the results of the toxicological analysis of the CRYIA(b) protein, the NPTII enzyme, and of transgenic Bt-tomato lines.

Development of Food Safety Test Strategy of Bt-tomatoes. The strategy for food safety evaluation of the transgenic Bt-tomato has been designed in a way that a number of pertinent questions can be answered: (i) does the CRYIA(b) protein exert a similar toxic action in mammals as observed in target insect species. It has been shown that the insecticidal activity of CRYIA(b) protein is mediated through binding to membrane receptors on epithelial cells of the insect midgut, disturbing the potassium/sodium membrane gradients, which leads to cellular swelling, vacuolization, and lysis of midgut epithelial cells (10-13); (ii) do the newly introduced proteins cause systemic adverse effects in mammals, in particular immunotoxic (allergenic) effects, and if so, can a No Effect Level be established, and; (iii) does the applied recombinant DNA technology lead to significant changes in the nutritional composition of the genetically modified tomato, or in the content of naturally occurring toxicants, which could negatively influence the safety of the product as food. A number of *in vivo* and *in vitro* toxicological and analytical experiments have been carried out in order to answer the questions raised.

Transformation and Characterization of Bt-Tomato Lines. Transgenic tomato plants were obtained from two different parental lines (TL001 a round type tomato, and SM002 a cylindrical elongated type tomato) by *Agrobacterium tumefaciens* (*A. tumefaciens*) mediated transformation using cotyledons as the explant. A disarmed *A. tumefaciens* strain containing the vector pPSO216 was used for transformation.

The vector Ppso216 comprises two chimeric genes between the T-DNA border repeats of the Ti plasmid. The chimeric *neo* gene consists of the promoter of the T-DNA TR1' gene (PTR1') (*14*), the coding region of the *neo* gene encoding neomycin phosphotransferase II (NPTII) from transposon Tn5 (*15*) and the 3' untranslated end from the octopine synthase gene (3'ocs) (*16*). The construction of the chimeric pnos-*neo*-3'ocs gene has been described by Hain et al. (*17*). The chimeric *B. thuringiensis* gene consists of the wound stimulated promoter of the T-DNA TR2' gene (PTR2') (*14*), the coding region of the C-terminal truncated *Bt*2 gene, called *IAb*6 derived from the coding region bt884 (*18*), and the 3' untranslated end from the T-DNA gene 7 (*19,20*). Based on entomological and agronomical criteria two transformants were chosen for further characterizations: the transgenic tomato variety RLE13-0009 derived from the parental line TL001, and the transgenic tomato variety RLE6-10001 derived from the parental line SM002.

Expression levels of the CRYIA(b) protein in transgenic Bt-tomatoes were tested in freshly harvested fruit (i.e. non-induced levels), and in extracts of sliced Bt-tomatoes which were incubated overnight in protoplast medium containing 1 mg/l 2,4-dichlorophenoxyacetic acid (2,4-D) (i.e. induced levels). An enzyme-linked-immunosorbent assay (ELISA) was used to determine the CRYIA(b) protein in tomato matrices (*21*). Levels of enzymatically active NPTII protein in ripe red tomatoes and processed products were quantified using Western blot analysis and dot blot assay in which the phosphorylation of kanamycin by ^{32}P-labelled ATP was monitored (*22*). In the harvested transgenic tomato variety RLE 13-0009 the CRYIA(b) protein was typically expressed in fresh fruit (i.e. non-induced) at levels of about 7.5 ng/mg protein, and in induced tomato fruit at levels of about 25.4 ng/mg of protein with a nominal content of 0.8% protein of fresh weight tomatoes.

Production and Characterization of CRYIA(b) and NPTII Protein. As recombinant DNA protein cannot be extracted easily from transgenic plant tissue, the coding sequences of the proteins CRYIA(b) and NPTII were introduced into *Escherichia coli* (*E. coli*) and fermentations were carried out to produce sufficient quantities for toxicology and binding studies as reported below.

To produce milligram quantities of purified CRYIA(b) protein, the *Bt*2 gene was cloned from *B. thuringiensis* var. *berliner* 1715 (*18,21*) which encodes the 130 kDa δ-endotoxin (*Bt*2-protoxin). After purification and solubilization from the *E. coli* strain K514 (pGI502) the protoxin was digested with trypsin and chymotrypsin to yield the active toxic protein CRYIA(b) with a relative molecular weight (Mr) of 66-68 kDa. Subsequent purification and characterization of the CRYIA(b) protein was carried out as described previously (*21*).

The *neo* gene, encoding the 264 amino acid protein NPTII, was isolated from the bacterial transposon Tn5 (*15*). The *neo* gene was cloned behind the lambda promoter to enable high level expression of the NPTII protein in *E. coli* (*23*). This construction resulted in the production of inclusion bodies containing the *cro*-NPTII protein, which precipitated selectively in the culture medium in milligram quantities with sufficient purity.

The Insecticidal Activity of CryIA(b) Protein in Mammals. The CRYIA(b) protein possesses enterotoxic effects on larvae of *Lepidoptera* insects such as *Heliothis armigera*, *Manduca sexta*, *Heliothis viriscens*, and *Phtorimaea opercullel-*

la. The CRYIA(b) protein appears to exert its toxic action in larvae of target insects, through specific interactions with the brush border membrane of the midgut epithelial cells (*21,24,25*). Although a quantitative correlation between the binding characteristics of a CRY protein and its insecticidal activity has not been reported, binding seems essential to the onset of toxicity (*26,27*). Therefore, the *in vivo* and *in vitro* binding of CRYIA(b) protein to gastro-intestinal (G.I.)-tract tissue sections of mammals has been investigated.

In vivo **Binding of CRYIA(b) Protein to G.I.-Tract Tissues of the Rat.** Male Brown Norway rats (n=4) were fed 0.5 mg of CRYIA(b) protein in 5 g of standard feed, corresponding to a human daily consumption of approximately 2000 kg of transgenic Bt-tomatoes, assuming a level of 60 µg CRYIA(b)/kg of fresh tomatoes (line RLE13-0009). Seven hours after administration the animals were sacrified for immunocytochemical analysis of the CRYIA(b) protein binding to tissues of esophagus, stomach, and intestine. The immunocytochemical approach used has been previously reported in studies with larvae of *Manduca sexta*, *Plutella xylostella* and *Leptinotarsa decemlineata* insects (*28*). For the detection of CRYIA(b) protein a polyclonal antiserum was used according to Höfte et al. (*18*). No binding of the CRYIA(b) protein could be observed in tissue segments taken from the esophagus, stomach, duodenum, ileum, jejunum, and colon. Moreover, no histopathological damage could be detected in G.I.-tract derived tissues from treated animals.

In vitro **Binding of CRYIA(b) Protein to G.I.-Tract Tissues of Mammals, including Men.** The *in vitro* binding of CRYIA(b) protein to intestinal tissue sections of rats, mice, rhesus monkey, and humans has been studied by modifying the immunocytochemical method as described by Bravo et al. (*29*). Briefly, deparafinated and hydrated tissue sections were incubated in Lugol (0.5% (w/w) I_2, 1% (w/w) KI and 5% (w/w) sodium thiosulphate). Subsequently, the washed tissue sections were incubated in methanol containing 0.6% (v/v) H_2O_2 for inactivation of endogenous peroxidase activity. After rinsing, the tissue sections were treated with a CRYIA(b) protein solution (10 µg/ml) at room temperature. The slides were then incubated overnight with mono- or polyclonal antiserum raised against the CRYIA(b) protein. The immunocytochemical analysis of CRYIA(b) protein in intestinal tissue sections of rat, rhesus monkey, and man was carried out using the corresponding monoclonal antiserum (mouse Ig). It appeared necessary to use a polyclonal antibody raised against CRYIA(b) (rabbit Ig) in mice tissues, in order to avoid cross reaction of the second antibody to IgA present in murine tissues of the small intestine. Control experiments were performed in the brush border epithelial cells of the midgut of the target insect larvae of *Manduca sexta*. In epithelial cell layers of segments of esophagus, stomach, duodenum, jejunum, ileum, and colon tissue of rats and mice, no specific binding of CRYIA(b) protein was observed, in contrast to uniform binding over the entire length of the brush border epithelium of the midgut of larvae of *Manduca sexta*. Similar experiments have been performed with intestinal tissues of rhesus monkey and humans, where no specific binding of CRYIA(b) protein could be observed. Except, a positive staining observed in caecum and colon tissue of the rhesus monkey, which could be reduced, however, by inhibition of the endogenous peroxidase activity using a prolonged incubation with methanol containing 0.6% (v/v) H_2O_2, indicating an aspecific effect.

General Toxicity Profile of CRYIA(b) Protein. Previous studies of the histopathology and mode of action of CRY proteins on mammals and other non-target species are very limited in number and have all used various crude spore-crystal mixtures of *B. thuringiensis* strains (*11,21,30,31*). Many *B. thuringiensis* strains produce more than one CRY protein, which complicates the elucidation of structure-toxicity relationships of the CRY proteins. On the other hand, the CRYIA(b) protein representing the domain of the *Bt*2 protoxin, that determines insect specificity and toxicity, has not yet been subjected to a detailed safety analysis. Tests were done with single doses of the CRYIA(b) protein to determine the digestibility and acute toxicity, as it has been shown that some *B. thuringiensis* toxins can resist proteases (*32*). To test for systemic effects of the CRYIA(b) protein upon passage through the intestinal wall, and for adverse immune reactions, daily doses of CRYIA(b) protein were fed to rodents for 30 days.

In vitro **Degradation of the Proteins CRYIA(b) and NPTII.** The degradation of CRYIA(b) protein, M_r 66-68 kDa, and of NPTII enzyme, M_r 27 kDa, was studied under simulated human gastro-intestinal conditions (*33*). Incubation of the proteins under conditions of pH 2, in the presence of pepsin (1 : 100, w/w), and subsequently at pH 8, in the presence of chymotrypsin and trypsin (1 : 25, w/w), revealed upon analysis by gel permeation HPLC and SDS-polyacryl-amide gel electrophoresis followed by immunoblotting, an extensive fragmentation of the proteins to peptides with molecular weights below 10 kDa. However, the degradation of the CRYIA(b) protein is clearly a two-step process. After two hours at pH 2 in the presence of pepsin the protein was readily cleaved to yield a 15 kDa fragment, and successively to smaller fragments ($<<$ 10 kDa) after continued treatment with chymotrypsin and trypsin at pH 8. The NPTII protein on the contrary appeared completely digested at pH 2 in the presence of pepsin.

Digestibility of CRYIA(b) Protein upon G.I.-Tract Passage in Rats. The digestibility of CRYIA(b) protein was studied in male Brown Norway rats (n=5), which were fistulated in the ileum before the caecum. Animals were fed 0.5 mg of CRYIA(b) protein in 5 g of standard feed, corresponding to an approximate human daily consumption of 2000 kg of the transgenic Bt-tomato line RLE13-0009. Chyme was collected after 5-7 hours, and immunoblot analysis revealed no intact CRYIA(b) protein, but fragments of 20-30 kDa and smaller. In samples taken after 7 hours no large fragments of CRYIA(b) protein were visible, indicating a further extensive degradation of the protein during gastro-intestinal tract passage into smaller peptides with molecular weights $<<$ 9 kDa.

Short-term Toxicity Study with CRYIA(b) Protein in Mice. The CRYIA(b) protein was orally dosed to female NMRI mice (n=10/group per dose) via drinking water *ad libidum*, at dose levels of 1.5 and 15 µg of CRYIA(b)/animal/day during 28 days. The highest dose level corresponded to a daily consumption of approximately 500 kg of the transgenic Bt-tomato variety RLE13-0009. Control mice were administered tap water, and test solutions were refreshed every third day. There were no differences observed in body weight gain, absolute and relative weights of liver and kidneys, and selected haematological parameters, including white blood cell differentional counts, between treated and control animals. Furthermore, histopathological analysis of the gastro-intestinal tract of animals treated with the CRYIA(b) protein during 28 days, did not reveal any signs of harmful effects induced by this insecticidal crystal protein from *B. thuringiensis*.

Short-Term Toxicity Study with CRYIA(b) Protein in Rabbits. New Zealand white male rabbits (n=4/group per dose) were dosed via drinking water *ad libidum* at a concentration of 0.75 mg CRYIA(b) protein/L during 31 days. Controls received tap water and test solutions were refreshed each day. Data of average water consumption indicated a daily intake of CRYIA(b) protein of 60 µg/kg of body weight, corresponding to a human daily consumption of approximately 60 kg of the transgenic Bt-tomato variety RLE13-0009. There were no changes observed in food consumption, and water intake nor in body weight gain and absolute and relative liver and kidney weights between treated and non-treated animals. No differences were noted in haematological parameters, including white blood cell differential counts. Furthermore, histological analysis of various segments of the gastro-intestinal tract did not reveal any harmful effect in the treated animals. Analysis of serum samples taken 2 and 4 weeks after initiation of the experiment did not indicate that antibodies against the CRYIA(b) protein were induced in treated animals. In addition, no significant differences were found in the total immunoglobulin (IgG) content of the serum of treated animals compared to that of control animals.

Haemolytic Effects of CRYIA(b) Protein. Human red blood cells (RBC's) were tested for the haemolytic potential of the CRYIA(b) protein, by incubation of the RBC's with the protein, and monitoring the osmotic fragility by electron microscopy and spectroscopically. No haemolysis was observed (Figure 1). Since it has been postulated that the site of interaction of the CRYIA(b) protein may be the enzyme ATPase, which is located on the cyto-plasmatic side of the membrane (*30*), the CRYIA(b) protein has been entrapped inside the RBC by the method of hypotonic dialysis (*34*), and subsequently tested for its haemolytic potency. Erythrocytes containing the CRYIA(b) protein showed negligible haemolysis comparable to that observed with RBC's swollen in the absence of CRYIA(b) protein or with RBC's containing entrapped albumin.

Food Safety of Transgenic Bt-Tomatoes. Resides studies of the recDNA CRYIA(b) protein in order to characterize the specific and systemic toxic potency of the protein, experiments have been focused on the safety of the whole transformed Bt-tomatoes. Genetic modification of tomato plants may induce changes in metabolic functions, which may lead to changes in the content of nutritional components and of naturally occurring glycoalkaloids. Such changes may affect the nutritional and/or toxicological characteristics of the new food product. Therefore, chemical analyses were made of a number of macro- and micronutrients, and of the most common glycoalkaloid in tomato, α-tomatine. Moreover, a 91 day feeding study was conducted with rats using diets containing 10% (w/w) of lyophilized transgenic Bt-tomatoes and control plants.

Chemical Analysis of Nutrients and Glycoalkaloids. After field testing the mature red tomatoes were harvested, and representative samples of the whole lot of freeze-dried tomatoes were analyzed for major nutrients under conditions corresponding to the expected time for processing (i.e. soon after harvest). Selected macro- and micronutrients of the transgenic Bt-tomato variety RLE13-0009 were compared to the nutritional composition of its respective nontransformed control line TL001 (Table I).

Table I. Nutritional Components in Lyophilized Ripe Fruit for Transgenic Bt-Tomatoes as Compared to the Control Variety and Normal Ranges

Nutrient	Control variety TL0001[b]	Transgenic variety RLE 13-0009[b]	Normal range[a]	Unit
Crude protein	130.4	148.2	104-208	g/kg[c]
Crude fat	20.3	17.2	30-63	g/kg
Carbohydrates	436	475	283-833	g/kg
Crude fiber	127	134	91-175	g/kg
Vitamin C	549	861	3448-6000	mg/kg
Calcium	2400	2610	1515-4375	mg/kg
Phosphorus	4120	4070	2652-5938	mg/kg
Chloride	6000	5100	3636-14375	mg/kg
Sodium	323	338	394-2083	mg/kg
Potassium	38200	39400	40000-65417	mg/kg
Magnesium	1620	1700	3030	mg/kg
Iron	82.2	151	61-125	mg/kg

[a]SOURCE: Adapted from ref. *35*.
[b]The chemical analysis of the final output was made on the whole lot of lyophilized tomatoes (about 160 kg of fresh tomatoes/batch per variety).
[c]Data are expressed as g or mg/kg of lyophilized tomatoes.

The values for the transformed Bt-tomatoes were within the typical ranges determined for nontransformed tomatoes, and found to be within published ranges (*35*). Also the chemical analysis of the transformed variety RLE6-10001 did not reveal significant differences in nutrients compared to the parental line SM002 (unpublished results). Thus, these chemical analyses indicated that transgenic Bt-tomatoes do not show significant changes in nutritional values as compared to non-transformed control tomatoes. No significant differences were observed in the concentrations of α-tomatine between mature red tomatoes of modified plants and controls (range: 1.4 - 1.7 mg α-tomatine/kg fresh weight of tomatoes). Other known solanaceous alkaloids (i.e. solanine) were not detected. However, the extraction procedures developed to isolate and quantitate the glycoalkaloids in potatoes (*36*) did not prove to be adaptable for use in tomatoes and had to be modified.

91-Day Feeding Trial with Transgenic Tomatoes. The field tested transgenic Bt-tomato variety RLE13-0009, and the respective control line TL001, were selected for a 91 day feeding study in rats. The field trial manifested no significant differences in vegetative growth and harvest characteristics between transgenic Bt-tomatoes and the controls. In the harvested transgenic tomato variety RLE13-0009 the CRYIA(b) protein was typically expressed in fresh tomatoes (i.e. under non-induced conditions) at levels of about 7.5 ng/mg protein, and in lyophilized tomatoes at levels of about 40.6 ng/mg of protein with a nominal protein content of 0.8% of fresh weight.

Previous range-finding studies showed that up to 10% (w/w) of lyophilized tomato fruit was well tolerated in the animal diet (*37*). However, insufficient data on the amino acid composition of tomato proteins, together with the observation of increased potassium urinary concentrations at day 30 in the 13 week dosed range-finding study, and the relatively high levels of carotene, limited further dose escalation. Therefore, three groups of 12 male and 12 female weanling Wistar rats were fed during a period of 91 days, respectively, a control semi-synthetic animal diet (Muracon SSP TOX), the same diet supplemented with 10% (w/w) of lyophilized transgenic Bt-tomato material of the variety RLE13-0009, or with 10% (w/w) of lyophilized material from the control parent line TL001. The macro- and micronutrient composition was equalized in all diets: 7.5% E (energy) fat, 20% E protein, 41.5% E carbohydrate and 10.4% fibre (w/w). The amounts of supplementary minerals and vitamins (i.e. vitamin C) were made allowance for the actual levels in lyophilized tomatoes (Table I). Survival, clinical signs, body weight, feed consumption, feed efficiency, absolute and relative organ weights, haematological values, and clinical chemistry parameters were measured and post-mortem analysis of tissues was performed. The average daily intake of tomato powder over the 91 day period corresponded to approximately 200 g of tomatoes/kg body weight, equivalent to a daily human consumption of 13 kg transgenic Bt-tomatoes. There was no feed refusal or unusual behavior in any of the animals. No significant differences in the other parameters under study have been noticed between the different diet groups, and no macroscopic abnormalities were found. The weights of liver, kidneys, spleen, and thymus, expressed as percent of body weight, summarized in Figure 2, did not show differences in treated animals compared to controls. The Ig-antibody measurements in serum and microscopic histological analysis included in this study are in progress.

Conclusions

Results obtained from the experiments with rats orally dosed with recDNA CRYIA(b) protein indicated that no specific receptors for the protein are present along the G.I.-tract, in contrast to what has been observed in the midgut of larvae of *Lepidoptera* insects such as *Manduca sexta* (*11,13*). Furthermore, no histopathological effects of the protein could be observed in the digestive mucosa cells, while in the midgut tissue of the target insects severe damage was noticed, i.e., vacuolization, profound swelling, and lysis of the mucosa cells. Moreover, binding assays *in vitro*, incubating CRYIA(b) protein with G.I.-tract tissue segments of rats, mice, rhesus monkey, and humans, did not reveal the presence of specific receptors

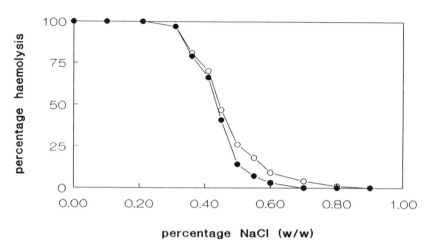

percentage NaCl (w/w)

Figure 1. Comparison of haemolysis in different NaCl solutions (% NaCl, weight by weight) between (closed circles) human red blood cells (RBC's) in the presence of 0.5 mg/ml CRYIA(b) protein (M_r 66-68 kDa) and (open circles) human RBC's swollen in the absence of CRYIA(b) protein according to the method of Zolla et al. (34). Washed RBC's were mixed at a final haematocrit of 70% and a protein/RBC ratio of 0.038 mM/1.10^6 cells.

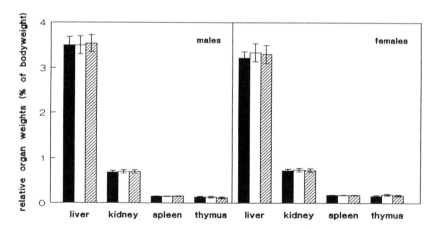

Figure 2. Organ weights in percent of body weight of male (left panel) and female (right panel) Wistar rats at the end of the 91 day feeding study: (solid bars) controls; (open bars) nontransformed-tomato fed animals; (hatched bars) transgenic Bt-tomato fed animals, mean ± SD.

for the CRYIA(b) protein. The CRYIA(b) protein and the NPTII marker protein degraded rapidly under simulated gastro-intestinal conditions to smaller fragments with molecular weights below 10 kDa, and CRYIA(b) protein is upon high dosage oral feeding to rats extensively digested in the G.I.-tract to smaller peptides.

The short-term feeding studies of CRYIA(b) protein administered to mice and rabbits, have been designed as pilot experiments, limited in the number of animals and in the experimental parameters examined. Relatively high dosages of the recDNA protein have been applied during these studies, corresponding to a daily human consumption of fresh transgenic Bt-tomatoes of 500 and 50 kg, respectively. These experiments did not reveal any signs of systemic adverse effects. No indications were found for specific immunotoxic effects of the CRYIA(b) protein as judged from the histological examination of spleen, lymph nodes, and the Peyer's patches of treated animals. Furthermore, in serum of treated rabbits no specific antibodies against CRYIA(b) protein could be detected, nor was the total IgG concentration elevated with respect to control animals. *In vitro* experiments to test for the haemolytic potency of the CRYIA(b) protein yielded negative results.

The question whether genetic manipulation may result in secondary changes in the metabolism of the transformed plant which could be of toxicological significance, has been tackled by two approaches. First, chemical analysis has been performed on the nutritional composition of the (un)modified lyophilized tomatoes, and second, a 91 day feeding trial with rats has been carried out in order to establish the food safety of the modified product. Chemical analysis did not reveal major changes in the content of macro- and micronutrients in transgenic Bt-tomatoes compared to those in the respective control line and to that of conventionally bred cultivars (*35*). It is important to note that nutritional components normally vary due to both cultivar-related and environmental influences (*38*). Moreover, levels of the glycoalkaloid, α-tomatine, were similar in modified and control tomatoes.

Preliminary analysis of the results of the 91 day feeding trial with rats, orally dosed with lyophilized tomato powder derived from a transgenic or a conventionally bred tomato variety, did not reveal any signs of adverse effects. The estimated average intake of tomato powder during the test period corresponded to a daily human consumption of 13 kg of fresh transgenic Bt-tomatoes. Food intake, body and organ weights, and clinical parameters were normal and gross macroscopic examination of tissues did not indicate toxic effects. Histopathological analysis of organs and tissues is in progress. Although final conclusions cannot be drawn yet, the toxicological data obtained up till now are reassuring with respect to food safety of the genetically modified Bt-tomatoes.

A number of aspects concerning the food safety of transgenic Bt-tomatoes still have to be studied:

(i) posttranslational modifications; studies performed up til now have been carried out with recDNA CRYIA(b) protein purified from *E. coli*. Differences in post translation modifications in prokaryotic and eukaryotic systems (i.e. degree and kind of amidation, glycosylation, phosphorylation of proteins), may influence the toxic potential of introduced proteins.

(ii) allergenic potency; the allergenic potency of the CRYIA(b) protein has

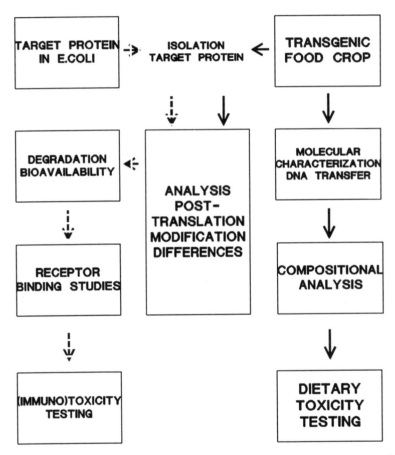

Figure 3. Proposed testing strategy for the safety assessment of transgenic food crops.

not been tested. Although long term use of *B. thuringiensis* crystal protein inclusions as a spray has not revealed evidence for allergic reactions in workers, studies on the allergenic potency of the CRYIA(b) protein should be considered, in particular, because of the potential general exposure of the population to CRY proteins via transgenic crop plants. As yet, however, there is a lack of predictive and/or validated *in vivo* models for the determination of allergenic potential of foods.

(iii) secondary effects; as generally recognized feeding test animals whole plant products suffers from a number of distinct disadvantages like relatively low exposure levels of compounds under study due to adverse effects from multicomponent exposure, naturally occurring toxicants, or nutrient imbalances, and thus, limited ranges of safety factors to be applied due to the nutritional or toxicological dietary restrictions. As a result, the latter type of experiments will play a diminishing role in the safety assessment of genetically modified crop plants and derived foods. On the other hand, chemical analysis of single nutrients and of naturally occurring toxicants has also its limitations, due to the lack of both appropriate analytical methods and knowledge about the composition of traditional foods in order to make comparisons effectively (*39*). Therefore, new ways of analysis should be explored, focusing on the characterization of whole crop or plant extracts with respect to profiles of secondary metabolites (i.e 'metabolic finger-printing') rather than on isolation and structural characterization of single compounds. To this end various analytical techniques for multicomponent analysis could be applied like HPLC, Capillary Zone Electrophoresis (CZE), NMR and LC(-GC)-MS(-MS). However, it should be emphasized that the identification of possible differences in recognition patterns depends to a considerable extent on the background of the natural variation in composition of food plants. Nevertheless, this analytical approach may offer possibilities to refrain from animal feeding trials with whole foods.

Risk evaluation of transgenic food crops with respect to human consumption must be based on: (i) the specific genetic modifications involved; (ii) the characteristics of introduced genes and expression products; (iii) possible secondary metabolic changes occurring as a result of gene manipulation, and (iv) levels of exposure to be expected in various foods. In Figure 3 a general scheme for testing of transgenic food crops is proposed. It is emphasized that safety assessment of genetically modified foods should take place within *a general framework* with *'case-by-case' variations*. The challenge for modern toxicology is to develop such flexible and tailormade test strategies for transgenic food crops, taking the often long history of safe use of traditional 'counterpart' foods into account. New analytical and *in vitro* toxicological approaches offer great possibilities to achieve these objectives.

Acknowledgments

We thank Dr. B. Verachtert of Plant Genetic Systems for excellent scientific collaboration. We thank M. Peters and G. van Tintelen for their assistance at the Centre for Small Laboratory Animals of the Agricultural University Wageningen, Dr. M.J. Groot, J.S. Ossekoppele for the gross examination. J.F. Labrijn, G.J.M.

Loeffen, G.D. van Bruchem, H.J. van Egmond, A. de Koning, A.R.M. Hamers, M.B.M. Huveneers-Oorsprong, and H.M. van de Putte are acknowledged for, respectively, histological, clinical, and chemical analysis. This work was carried out in the framework of the EU Food-Linked Agro-Industrial Research (FLAIR) program (contract no. AGRF-CT90-0039).

Literature Cited

1. Fraley, R. *Bio/Technology* **1992**, *10*, 40-43.
2. Gadani, F.; Mansky, L.M.; Medici, R.; Miller, W.A.; Hill, J.H. *Arch. Virol.* **1990**, *115*, 1-21.
3. Jones, J.L. *Trends in Food Science & Technology* **1992**, *3*, 54-59.
4. Lindsey, K. *Agro-food-Industry, Hi-tech* **1991**, *4*, 8-16.
5. Oxtoby, E. *Tibtech* **1990**, *8*, 61-65.
6. Perlak, F.J.; Fuchs, R.L.; Dean, D.A.; McPherson, S.L.; Fischhoff, D.A. *Proc. Natl. Acad. Sci. USA* **1991**, *88*, 3324-3328.
7. Kok, E.J.; Reynaerts, A.; Kuiper, H.A. *Trends in Food Science & Technology* **1993**, *4*, 42-48.
8. Höfte, H.; Whitely, H.R. *Microbiol. Rev.* **1989**, *53*, 242-255.
9. Adang, M.; DeBoer, D.; Endres, J.; Firoozabady, E.; Klein, J.; Merlo, D.; Murray, E.; Rashka, K.; Stock, C. In *Biotechnology, Biological Pesticides and Novel Plant-Pest Resistance for Insect Test Management*; Roberts, D.W.; Granados, R.R., Eds.; Insect Pathology Resource Center, Boyce Thompson Institute for Plant Research at Cornell University, Ithaca, NY, **1989**; pp 31-37.
10. Hendrickx, K.; De Loof, A.; Van Mellaert, H. *Comp. Biochem. Physiol.* **1990**, *95C*(2), 241-245.
11. Knowles, B.H.; Ellar, D.J. *Biochim. Biophys. Acta* **1987**, *924*, 509-518.
12. Endo, Y.; Nishiitsutsuji-Uwo, J. *J. Invertebr. Pathol.* **1980**, *36*, 90-103.
13. Percy, J.; Fast, P.G. *J. Invertebr. Pathol.* **1983**, *41*, 86-98
14. Velten, J.; Velten, L.; Hain, R.; Schell, J. *EMBO J.* **1984**, *3*, 2723-2730.
15. Beck, E.; Ludwig, G.; Auerswald, E.; Reiss, B.; Schaller, H. *Gene* **1982**, *41*, 327-336.
16. Gielen, J.; De Beuckeleer, M.; Seurinck, J.; Deboeck, F.; De Greve, H.; Lemmers, M.; Van Montagu, M.; Schell, J. *EMBO J.* **1984**, *3*, 835-846.
17. Hain, R.; Stabel, P.; Czernilofsky, A.P.; Steinbiss, H.H.; Herrera-Estrella, L.; Schell, J. *Mol. Gen. Genet.* **1985**, *199*, 161-168.
18. Höfte., H.; De Greve, H.; Seurinck, J.; Jansens, S.; Mahillon, J.; Ampe, C.; Vandekerckhove, J.; Vanderbruggen, H.; Van Montagu, M.; Zabeau, M.; Vaeck, M. *Eur. J. Biochem.* **1986**, *161*, 273-280.
19. Velten, J.; Schell, J. *Nucleic Acids Res.* **1985**, *13*, 6981-6998.
20. Dhaese, P.; De Greve, H.; Gielen, J.; Seurinck, J.; Van Montagu, M.; Schell, J. *EMBO J.* **1983**, *2*, 419-426.
21. Hoffmann, C.; Vanderbruggen, H.; Höfte, H.; Van Rie, J.; Jansens, S.; Van Mellaert, H. *Proc. Natl. Acad. Sci. USA* **1988**, *85*, 7844-7848.

22. Radke, S.E.; Andrews, B.M.; Moloney, M.M.; Crouch, M.L., Kridl, J.C.; Knauf, V.C. *Theor. Appl. Genet.* **1988**, *75*, 685-694.
23. Botterman, J.; Höfte, H.; Zabeau, M. *J. Biotechnology* **1987**, *6*, 71-81.
24. Van Rie, J.; McGaughey, W.H.; Johnson, D.E.; Barnett, B.D.; Van Mellaert, H. *Science* **1990**, *2476*, 72-74.
25. Van Rie, J.; Jansens, S.; Höfte, H.; Degheele, D.; Van Mellaert, H. *Appl. Environ. Microbiol.* **1990**, *56*, 1378-1385.
26. Wolfersberger, M.G. *Experientia* **1990**, *46*, 475-477.
27. Ferré, J.; Real, M.D.; Van Rie, J.; Jansens, S.; Peferoen, M. *Proc. Natl. Acad. Sci. USA* **1991**, *88*(12), 5119-5123.
28. Bravo, A.; Hendrickx, K.; Jansens, S.; Peferoen, M. *J. Invertebr. Pathol.* **1992**, *60*(3), 247-253.
29. Bravo, A.; Jansens, S.; Peferoen, M. *J. Invertebr. Pathol.* **1991**, *60*(3), 237-246.
30. English, L.H.; Cantley, L.C. *J. Biol. Chem.* **1986**, *261*, 1170-1173.
31. Nishiitsutsuji-Uwo, J.; Endo, Y.; Himeno, M. *Appl. Ent. Zool.* **1980**, *15*, 133-139.
32. Bietlot, H.; Carey, P.R.; Choma, C.; Kaplan, H.; Lessard, T.; Pozgav, M. *Biochem. J.* **1989**, *260*, 87-91.
33. Gauthier, S.F.; Vachon, C.; Savoie, L. *J. Food Sci.* **1986**, *51*, 960-964.
34. Zolla, L.; Lupidi, G.; Marcheggiani, M.; Falcioni, G.; Brunori, M. *Biochim. Biophys. Acta* **1990**, *1024*, 1-9.
35. *Food Composition and Nutrition Tables 1986/87*; Souci, S.W.; Fachmann, W.; Kraut, H., Eds.; Wissenschaftliche Verlagsgesellschaft mbH, Stuttgart, **1986**, pp 694-695.
36. Van Gelder, W.M.J.; Vinke, H.; Scheffer, J.J.C. *Euphytica* **1988**, *39*, 147-158.
37. Noteborn, H.P.J.M.; Bienenmann-Ploum, M.E.; Groot, M.J.; Alink, G.M.; Reynaerts, A.; Malgarini, G.; Kuiper, H.A. *Med. Fac. Landbouww. Univ. Gent (MFLRA3)* **1994**, *59*(4a), 1765-1774.
38. Davies, J.N.; Hobson, G.E. *CRC Critical Reviews in Food Science and Nutrition* **1981**, *15*, 205-280.
39. Hall, R.L. *Food Technol.* **1992**, *46*(12), 109-112.

RECEIVED June 7, 1995

Chapter 13

Safety Assessment of Potatoes Resistant to Colorado Potato Beetle

P. B. Lavrik[1], D. E. Bartnicki[1,4], J. Feldman[2], B. G. Hammond[1],
P. J. Keck[1], S. L. Love[3], M. W. Naylor[1], G. J. Rogan[1], S. R. Sims[1],
and R. L. Fuchs[1]

[1]The Agricultural Group, Monsanto Company, Chesterfield, MO 63198
[2]Potato Products Division, Hybritech Seed International, Inc.,
Boise, ID 83706
[3]College of Agriculture, University of Idaho, Aberdeen, ID 83210

Russet Burbank potato plants which resist damage by Colorado potato beetle (CPB) have been developed through *Agrobacterium*-mediated transformation. The CPB resistant potato plants produce two additional proteins, the *Bacillus thuringiensis* subsp. *tenebrionis* protein which confers resistance to CPB and the neomycin phosphotransferase II protein which serves as a selectable marker during the plant transformation process. Extensive studies have been performed to assess the environmental and food safety of the CPB resistant potatoes to gain regulatory approval for commercialization of potato varieties with this new trait. Data derived from the studies have confirmed the safety of the newly introduced proteins and demonstrated that the presence of these proteins and the process used to produce these plants did not alter the quality, composition or safety of the tubers from CPB resistant plants as compared to commercial Russet Burbank potatoes.

Colorado potato beetle (CPB; *Leptinotarsa decemlineata*) is the most damaging insect pest of the $2.3 billion U.S. potato crop (*1,2*). If untreated or poorly managed, the CPB can devastate potato production in some areas and lower yield by as much as 85 percent (*3-5*). Current CPB control primarily involves the use of a diminishing number of broad spectrum insecticides. These insecticides are variably effective due to environmental factors and insect susceptibility, with seasonal control costs which can exceed $200 per acre (*6*). Application of new technologies to control this pest is essential for continuance of potato production in some areas.

[4]Current address: Calbiochem-Novachem, 10394 Pacific Center Court, San Diego, CA 92121

Through genetic modification, Russet Burbank potato plants have been produced which are virtually immune to damage by CPB (7). The CPB resistant plants were produced by inserting the gene coding for the CPB-active protein from *Bacillus thuringiensis* subsp. *tenebrionis* (*B.t.t.*) into the potato genome using *Agrobacterium tumefaciens*-mediated transformation (7).

Field trials conducted during the past four years with CPB resistant potatoes have demonstrated effective, season long control of feeding damage by all stages of the CPB, including the recalcitrant adult, without insecticide application (7). Agronomic evaluation, consisting of plant growth characteristics, general insect and disease susceptibility, yield and tuber quality have shown the CPB resistant potatoes to be indistinguishable from control Russet Burbank potatoes (7). The superior CPB control offered by the plants will enable growers to significantly reduce the amount of chemical insecticide now applied to their crop and promote increases in beneficial insect populations which can help reduce other potato pests not directly controlled by the CPB-active protein.

Safety Assessment of CPB Resistant Potatoes

As with other food crops improved through biotechnology, the United States Department of Agriculture (USDA), the Environmental Protection Agency (EPA) and the Food and Drug Administration (FDA) exercise joint regulatory oversight for these genetically improved potatoes. The USDA, under the jurisdiction of the Plant Pest Quarantine Act, regulates the impact of the genetically modified plant on the environment and production agriculture according to guidelines published in 1993 (8). The EPA, under the jurisdiction of the Federal Insecticide, Fungicide and Rodenticide Act (FIFRA), regulates the environmental and food safety, respectively, of both the pesticidally active (e.g. *B.t.t.* protein) and inert (e.g. NPTII protein) ingredient introduced into the plant according to proposed guidelines (9). The FDA, under the jurisdiction of the Federal Food, Drug and Cosmetic Act (FFDCA), oversees the safety and wholesomeness of the food and/or feed according to the new plant variety food policy published in 1992 (10). CPB resistant potatoes are the first genetically modified crop possessing a pesticidal trait to be reviewed for full approval by the regulatory agencies and can thereby serve as a case study for the overall safety assessment strategy (11).

The safety of the CPB resistant potatoes was assessed according to the published guidelines and policies of the three regulatory agencies. The concept of "substantial equivalence" (also referred to as "no material difference" – terms which encompass a demonstration that the new potato variety is comparable to traditionally bred potatoes in terms of nutritional quality, level of important natural products, and agronomic and environmental performance) was used to assess the safety of these potatoes (10,12). The studies focused on two primary items: (1) the safety of the additional expressed proteins to the environment and in food/feed, and (2) the composition and wholesomeness of the potato tuber. Safety assessments were conducted on seven independently transformed CPB resistant lines.

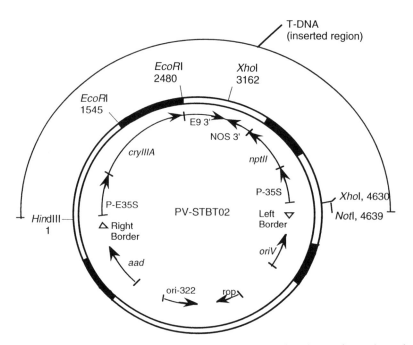

Figure 1.   Plasmid used in the *Agrobacterium*-mediated transformation of Russet Burbank potato plants.   T-DNA, region inserted into the potato genome, from the right border to the left border, the region contains: the E35S cauliflower mosaic virus promoter with the duplicated enhancer region; the *cryIIIA* gene which confers resistance to CPB; the E9 3' terminator of the pea ribulose-1,5-bisphosphate carboxylase; NOS 3' terminator from the nopaline synthase gene; the *nptII* gene which codes for neomycin phosphotransferase type II enzyme; and the 35S cauliflower mosaic virus promoter.   Regions of DNA present in the plasmid but not transfered into the potato genome: *aad,* the gene for the enzyme streptomycin adenylyl-transferase that allows for bacterial selection on spectinomycin or streptomycin; *ori-322/rop*, a segment of pBR322 which provides the origin of replication for maintenance of the plasmid in *E. coli* and the replication of primer for conjugational transfer into *Agrobacterium* cells; *oriV*, origin of replication for the *Agrobacterium. Eco*RI, *Hin*dIII, *Xho*I and *Not*I, endonucleases restriction sites and their location in base pairs.

Safety of Newly Expressed Proteins. The CPB resistant potato plants produce two additional proteins, the *B.t.t.* protein which confers protection to CPB and the neomycin phosphotransferase (NPTII) protein which enabled the selection of transformed potato cells expressing the *B.t.t.* protein in tissue culture. In assessing the safety of these newly introduced proteins in food/feed and to the environment, the safety of the donor organism as well as the potato transformation process was also considered.

Safety of Donor Organisms and the Transformation Process. Two genes were introduced into the Russet Burbank genome, the *cryIIIA* and the *nptII* genes. The *cryIIIA* gene *(13)* was obtained from *Bacillus thuringiensis* subsp. *tenebrionis* *(B.t.t.)* strain BI 256-82 isolated by Krieg *et al.* *(14)* and encodes the CPB-active *B.t.t.* protein. This bacterium belongs to a large group of common spore-forming soil bacteria that produce proteinaceous parasporal crystalline inclusions that are lethal to specific classes of insects *(15)*. For over 30 years, a number of species from this group of bacteria have been safely used as commercial microbial insecticide formulations *(16)*. Microbial formulations of *B.t.t.* have been in commerce since 1988 with an equal record of safe usage *(17,18)*. Based on the available scientific data, EPA and other regulatory agencies worldwide have determined that the use of registered *Bacillus thuringiensis* products, including *B.t.t.,* pose no significant risks to human health or non-target organisms *(19)*.

The *nptII* gene was obtained from the procaryotic transposon Tn5 *(20)* and it codes for the neomycin phosphotransferase II enzyme which serves as a selectable marker protein during the plant transformation process. Organisms containing this protein are ubiquitous in the environment and within mammalian digestive systems *(21)*. Many new plant varieties, improved through genetic engineering, utilize this enzyme as a selectable marker *(22)*. The safety of this protein has been well established and the USDA, FDA and EPA have approved its use in crops *(23-25)*.

Introduction of the *cryIIIA* and *nptII* genes into the genome of seven CPB resistant lines under development was mediated by *Agrobacterium tumefaciens* harboring the double border binary vector, PV-STBT02, schematically shown in Figure 1. Southern hybridization analysis *(26)* of the DNA from each of the seven lines confirmed that the *cryIIIA* and the *nptII* genes, delineated by the right- and left-borders (depicted as the T-DNA region in Figure 1), were the only genes inserted into the potato genome. The analysis also demonstrated that complete insert integrity (gene size, composition and linkage) was maintained during the transformation process.

Protein Expression Levels. The *B.t.t.* and NPTII proteins are expressed at extremely low levels in potato leaf and tuber tissues of all seven lines. Based on data obtained by validated enzyme linked immunosorbent assays (ELISA's) from plant tissue collected at seven field sites throughout the commercial potato growing regions of the United States, the expression level of the *B.t.t.* protein in the seven CPB resistant lines was estimated to range from 5 to 28 μg per gram fresh weight of leaf tissue (approximately 0.12% of the total leaf protein) and 0.4 to 2.0 μg per gram fresh weight of tuber tissue (< 0.01% of the total tuber protein). The expression level of the NPTII protein in both the leaf or tuber tissues was estimated to be approximately 2- to 5-fold lower than the expression of the *B.t.t.* protein. The expression of the *B.t.t.* and NPTII proteins in foliage remains relatively constant throughout the season with the expected decline at the onset of plant senescence.

Safety of the *B.t.t.* Protein. The amino acid sequence encoded by the *crylIIA* gene inserted into the potato plants produces a protein identical to that produced by the *B.t.t.* microbe found in nature and also present in registered commercial microbial pesticide formulations (*7,27-29*). This protein is selectively active against a narrow spectrum of Coleoptera without affecting non-target insects, humans or animals (30). Upon ingestion by susceptible insect species, the protein binds to specific receptors in the mid-gut, feeding is inhibited with disruption of the gut epithelium and eventual death of the insect pest (*31*). Since other non-target organisms, including other classes of insects as well as birds, fish, and mammals do not possess such receptors, the *B.t.t.* protein does not affect them (*32,33*).

Even though the *B.t.t.* protein produced in CPB resistant potatoes is the same protein contained in commercial microbial products, many of the previous studies were repeated to verify the safety and selectivity of this protein. However, in contrast to previous studies which assessed the safety of microbial formulations, the present studies utilized highly purified *B.t.t.* protein as the test material. Since limited expression of this protein prohibited the isolation of large quantities of this protein from the potato tubers or potato plant directly, gram quantities of the *B.t.t.* protein were produced in *E. coli* engineered with the same gene present in the potato plants. Minor amounts of this protein, purified from the potato tuber and from *E. coli*, were shown to be chemically and functionally equivalent by a series of commonly used analytical assays, which included molecular weight, N-terminal amino acid analysis, immuno-reactivity, lack of glycosylation and insecticidal potency.

An acute gavage study was conducted in mice to confirm the mammalian safety of the *B.t.t.* protein. Following EPA guidelines, a high dose of the test substance (5000 mg of *B.t.t.* protein per kg body weight) was administered to the test animals. The dose was equivalent to over a 2.5 million-fold safety factor based on the average human consumption of potatoes and the level of the *B.t.t.* protein present in the tuber. No adverse effects were observed in test animals and there were no effects on food consumption, weight gain, or gross pathology. Additionally, purified protein was subjected to an *in vitro* digestion experiment which demonstrated that the *B.t.t.* protein has an extremely short half-life (less than 30 seconds) under simulated human gastric conditions (*34*). These studies confirm the safety of the *B.t.t.* protein to humans and animals.

The specificity of the *B.t.t.* protein to CPB was confirmed by insect host-range studies, also using purified *B.t.t.* protein isolated from *E. coli*. Five non-target beneficial insects (including adult and larval honey bees, *Apis mellifera*; lacewings,

Chrysopa carnea; ladybird beetles, *Hippodamia convergens*; and a parasitic wasp, *Nasonia vitripennis*) were shown to be unaffected by high doses of the purified *B.t.t.* protein (greater than 100 times the lethal concentration to CPB). These studies confirm the specificity and the safety of the *B.t.t.* protein to nontarget insects.

As with most common proteins, the *B.t.t.* protein is not expected to persist in the environment. Studies were conducted on the loss of bioactivity of the *B.t.t.* protein in potato plant tissue incorporated into soil (at levels approximately 35-fold higher than that estimated under field conditions) and in plant tissue without soil. Under both conditions, complete decay of biological activity was noted within 9 days. Therefore, a rapid degradation of *B.t.t.* protein bioactivity in potato plant tissues remaining in the field after potato tuber harvest is predicted.

Safety of the NPTII Protein. The NPTII protein was used as a selectable marker to enable the identification of potato cells containing the *cryIIIA* gene. The safety assessment of the NPTII protein has been discussed in detail in the FDA approval of the protein as a processing aid food additive *(24,35)* and in recent articles by Fuchs *et al. (36,37)*. The latter articles describe acute gavage and digestive fate studies identical to those conducted for the *B.t.t.* protein, with similar results. No adverse affects were observed in a mouse acute gavage study in which mice were administered a dose (5000 mg/kg) of NPTII equivalent to 5 million-fold greater than the projected human consumption. Similar to the results observed with the *B.t.t.* protein, the NPTII protein was shown to rapidly degrade in the simulated digestive fate study, with a half-life of less than 20 seconds in gastric fluid. These studies confirm the mammalian safety of this protein.

Composition and Wholesomeness of the Potato Tuber. Safety assessment of the tuber focused on whether the CPB resistant potatoes are "substantially equivalent" to the parental Russet Burbank variety. These varieties were evaluated in respect to composition, nutritional quality and in aspects that could impact the use, value, the environment, or food/feed safety of this product.

In terms of nutrition, potato has long been valued as an excellent source of staple starch and contributes mainly to the carbohydrate calories of the diet. More recently, potato has also been valued for selected vitamins and minerals such as vitamin C, vitamin B_6, and potassium (National Potato Board). Tubers from CPB resistant potato plants were analyzed for these key components of interest as well as a number of other constituents. The compositional analyses were more extensive than the analyses carried out on new commercial potato varieties developed through classical breeding. The analyses included a total of 23 components: proximate composition (total protein, total carbohydrate, fat, dietary fiber and ash), vitamins (vitamin C, vitamin B_6, thiamine, niacin, folic acid and riboflavin), minerals (calcium, copper, iron, iodine, magnesium, phosphorous, potassium, sodium and zinc) and glycoalkaloids (solanines and chaconines), which are the important natural toxicants present in all potatoes. The levels of these constituents in each of the seven CPB resistant potato lines were comparable to the parental Russet Burbank control and

Table I. Composition of CPB Resistant and Parental Russet Burbank
Potatoes

Component	CPB Resistant[a] Mean	Range	Russet Burbank Control[a] Mean	Range	Published Range
Solids, % tuber fresh weight	19.6	18.0 - 21.0	20.0	19.6 - 20.5	16.8 - 24.5[b]
Carbohydrate, g/100g tuber	16.0	15.4 - 16.5	16.0	15.7 - 16.4	13.0 - 17.0[c]
Protein, g/100g tuber	2.1	2.1 - 2.2	2.1	2.0 - 2.1	1.4 - 2.9[c]
Vitamin C, mg/100g tuber	11.4	8.7 - 13.6	11.6	11.0 - 12.3	10.3 - 22.0[b]
Vitamin B_6, µg/100g tuber	97.2	75.4 - 119.0	97.2	89.2 - 105.0	140.0 - 280.0[c]
Folic acid, µg/100g tuber	6.7	5.7 - 7.7	7.0	5.2 - 8.7	4.0 - 20.0[c]
Potassium, mg/100g tuber	420	388 - 453	416	393 - 438	340 - 600[c]
Glycoalkaloids, mg/100g tuber	3.8	2.7 - 5.8	3.1	2.7 - 3.5	3.1 - 16.1[b]

[a]Values are the mean of tubers obtained from seven CPB resistant or parental Russet Burbank control plant lines grown at two field locations. At each field location, plots for each CPB resistant line were replicated six times.
[b]Taken from Russet Burbank tubers grown in Aberdeen, ID (Pavek, J. *et al. Western Regional Variety Trial Report.* 1980-1992, WRCC-27, University of Idaho, ID).
[c]Range of values for white potatoes, Scherz, H. and Senser, F. In *Food Composition and Nutrition Tables 1989/90,* Deutsche Forshungsanstalt für Lebensmittelchemie, Garching b. Müchen, Eds Wissenschaftliche Verlagsgesellschaft mbH, Stuttgart, 1989, pp 542-544.

within the range of values reported in the literature. A comparison of the key components is summarized in Table I.

To assess the overall wholesomeness of the CPB resistant potatoes, raw potato tubers from both the CPB resistant and Russet Burbank control plants were fed, along with the regular diet, to rodents in a 28-day study. In this study, rats consumed an average of 80 g of potatoes per kg of body weight per day, which is equivalent to a human consumption of 35 to 40 potato tubers per day. No differences in food consumption, growth rate, behavior, or gross pathology were observed during these studies.

Assessment of the Allergenic Potential. Since potatoes are rarely known to cause allergic reactions, the allergenic potential assessment of CPB resistant potatoes focused primarily on the potential allergenic concerns of the two introduced proteins, the *B.t.t.* and NPTII proteins. These proteins have a long history of safe use and do not share the biochemical profile common to known allergenic proteins. Therefore, it is unlikely these proteins would pose any significant allergenic concerns. An assessment of the potential allergenic concerns for NPTII protein was considered by the FDA in their approval of the use of this protein as a processing aid food additive and found to pose no significant allergenic concerns (*24*). As discussed above, the *B.t.t.* protein expressed in CPB resistant potato plants is indistinguishable from the *B.t.t.* protein contained in commercial microbial formulations that have been used safely for the past six years on a variety of crops including fresh produce like eggplant and tomato, with no reported allergenic responses, thus supporting the lack of allergenic concern for the *B.t.t.* protein. Although there are no assays available to definitively predict the allergenic potential of proteins, the biochemical profile of the *B.t.t.* protein provides a qualitative basis for allergenic assessment when compared to known protein allergens. In general, protein allergens are present at high concentrations in foods that elicit an allergenic response; are stable to peptic and tryptic digestion and the acid conditions of the digestive system; and are often glycosylated (*38-40*). The *B.t.t.* protein does not possess any of these characteristics. The *B.t.t.* protein is present in CPB resistant potatoes at extremely low amounts; it was shown to be very labile in the mammalian gastric digestive system, minimizing any potential for this protein to be absorbed by the intestinal mucosa; and, the *B.t.t.* protein, as purified from the tuber of CPB resistant plants, is not glycosylated. Furthermore, a comparison of the *B.t.t* protein sequence by either a FASTa- (*41*) or Monte Carlo-type (*42*) analysis showed no significant homology to any known protein allergens in protein data bases (Pir protein, Swissprot, and Genpept protein). Based on this qualitative assessment, we conclude that CPB resistant Russet Burbank potatoes pose no additional allergenic concerns than any other potato in commerce.

Conclusions

CPB resistant Russet Burbank potato plant lines generated through genetic modification were assessed for environmental and food/feed safety based on the safety of the introduced proteins and the concept of "substantial equivalence." A direct safety assessment of the newly introduced *B.t.t.* and NPTII proteins, confirmed the safety of these components. The CPB resistant potato plant lines were shown to be substantially equivalent to traditionally bred potatoes in terms of nutritional quality, level of important natural products, and agronomic and environmental performance. Substantial equivalence was established for seven independently transformed CPB resistant plant lines with the parental Russet Burbank line which demonstrates a generic lack of effect by introduction of two genes in the potato genome.

Literature Cited

1.   Casagrande, R. A. *Bull. Entomol. Soc.* **1987**, 33, 142-150.
2.   Potato Statistical Yearbook, National Potato Council, Englewood, CO, 1992.
3.   Hare, J. D. *J. Econ. Entomol.* **1980**, 73, 369-373.
4.   Ferro, D. N.; Morzuch, B. J.; Margolies, D. *J. Econ. Entomol.* **1983**, 76, 349-356.
5.   Shields, E. J.; Wyman, J. A. *J. Econ. Entomol.* **1984**, 77, 1194-1199.
6.   Ferro, D. N.; Boiteau, G. In *Plant Health Management in Potato Production;* Rowe, R. C., Ed.; Am. Phytopath. Soc. Press: St. Paul, MN, 1992, pp 209-234.
7.   Perlak, F. J.; Stone, T. B.; Muskopf, Y. M.; Petersen, L. J.; Parker, G. B.; McPherson, S. A.; Wyman, J.; Love, S.; Reed, G.; Biever, D.; Fischhoff, D. A. *Plant Molec. Biol.* **1993**, 22, 313-321.
8.   U. S. Department of Agriculture (USDA). *Federal Register* **1993**, 58, 17044-17059.
9.   U. S. Environmental Protection Agency (EPA). *Federal Register* **1992**, 57, 55531-55532.
10. U. S. Food and Drug Administration (FDA). *Federal Register* **1992**, 57, 22984-23005.
11. U. S. Environmental Protection Agency (EPA). *Federal Register* **1993**, 58, 64582-64583.
12. Organization for Economic Cooperation and Development (OECD). *Safety Evaluation of Foods Derived by Modern Biotechnology: Concepts and Principles,* Paris, 1992.
13. Höfte, H.; Whiteley, H. R. *Microbiol. Rev.* **1989**, 53, 242-255.
14. Krieg, A.; Huger, A. M.; Langenbruch, G. A.; Schnetter, W. *Z. Angew., Entomol.* **1983**, 96, 500-508.
15. Aronson, A. I.; Beckman, W.; Dunn, P. *Microbiol. Rev.* **1986**, 50, 1-24.

16. Frankenhuyzen, K. In *Bacillus thuringiensis, An Environmental Biopesticide: Theory and Practice;* Entwistle, P. F.; Cory, J. S.; Bailey, M. J.; Higgs, S., Eds.; John Wiley & Sons, New York, NY, 1993, pp 1-35.
17. U. S. Environmental Protection Agency (EPA). EPA/OPP Pesticide Fact Sheets Chemical Code Names 006405 and 128946-1.
18. Keller, B.; Langenbruch, G. A. In *Bacillus thuringiensis, An Environmental Biopesticide: Theory and Practice;* Entwistle, P. F.; Cory, J. S.; Bailey, M. J.; Higgs, S., Eds.; John Wiley & Sons, New York, NY, 1993, pp 171-191.
19. U. S. Environmental Protection Agency (EPA). *National Technical Information Service (NTIS) PB 89-164198;* U. S. Government Printing Office, Washington, DC, 1988.
20. Beck, E.; Ludwig, G.; Auerswald, E. A.; Reiss, B.; Schaller, H. *Gene* **1982**, 19, 327-336.
21. Nap, J. P.; Bijvoet, J., Stikema, W. J. *Transgenic Research* **1992**, 1, 239-249.
22. World Health Organization (WHO). *Health Aspects of Marker Genes in Genetically Modified Plants.* WHO Food Safety Unit Workshop Report, Copenhagen, Denmark, 1993, pp 1-32.
23. U. S. Department of Agriculture (USDA). *Federal Register* **1992**, 57, 47608-47616.
24. U. S. Food and Drug Administration (FDA). *Federal Register* **1994**, 59, 26700-26711.
25. U. S. Environmental Protection Agency (EPA). *Federal Register* **1994**, 59, 49351-49353.
26. Southern, E. M. *J. Mol. Biol.* **1975**, 98, 503-517.
27. McPherson, S. A.; Perlak, F. J.; Fuchs, R. L.; Marrone, P. G.; Lavrik, P. B.; Fischhoff, D. A. *Bio/Technology* **1988**, 6, 61-66.
28. Herrnstadt, C.; Gilroy, T. E.; Sobieski, D. A.; Bennett, B. D.; Gaertner, F. H. *Gene* **1987**, 57, 37-46.
29. Donovan, W. P.; Gonzalez Jr, J. M.; Gilbert, M. P.; Dankocsik, C. *Mol. Gen. Genet.* **1988**, 214, 365-372.
30. MacIntosh, S. C.; Stone, T. B.; Sims, S. R.; Hunst, P. L.; Greenplate, J. T.; Marrone, P. G.; Perlak, F. J.; Fischhoff, D. A.; Fuchs, R. L. *J. Invert. Path.* **1990**, 56, 258-266.
31. Slaney, A. C.; Robins, H. L.; English, L. *Insect Biochem. Molec. Biol.* **1992**, 22, 9-18.
32. Hofmann, C.; Vanderbruggen, H. V.; Höfte, H.; Van Rie, J.; Jansens, S.; Van Mellaert, H. *Proc. Natl. Acad. Sci. USA* **1988**, 85, 7844-7848.
33. Sacchi, V. F.; Parenti, P.; Hanozet, G. M.; Giordana, B.; Lüthy, P.; Wolfersberger, M. G. *FEBS Lett.* **1986**, 204, 213-218.
34. United States Pharmacopeia, United States Pharmacopeial Convention, Inc.: Washington, DC, Vol. XXII; pp 1788-1789.
35. Calgene, Inc. *FDA Docket Number: 90A-0416,* 1990.

36. Fuchs, R. L.; Heeren, R. A.; Gustafson, M. E.; Rogan, G. J.; Bartnicki, D. E.; Leimgruber, R. M.; Finn, R. F.; Hershman, A.; Berberich, S. A. *Bio/Technology,* **1993,** 11, 1537-1542.

37. Fuchs, R. L.; Ream, J. E.; Hammond. B. G.; Naylor, M. W.; Leimgruber, R. M.; Berberich, S. A. *Bio/Technology,* **1993,** 11, 1543-1547.

38. Taylor, S. L. *Food Technol.,* **1992,** 39, 146-152.

39. Taylor, S. L.; Nordlee, J. A.; Bush, R. K. In *Food Safety Assessment;* Finley, J. W.; Robinson, S. F.; Armstrong, D. J., Eds; ACS Symposium Series No. 484, American Chemical Society: Washington, D.C., 1992, pp 316-329.

40. Metcalf, D. D. *Clin. Rev. Allergy,* **1985,** 3, 331-349.

41. Pearson, W. R. *Meth. Enzymol.,* **1990,** 183, 63-98.

42. Needleman, S. B.; Wunsch, C.P. *J. Mol. Biol.,* **1970,** 48, 443-453.

RECEIVED June 7, 1995

MICROORGANISMS

Chapter 14

Improvement of Beer Brewing by Using Genetically Modified Yeast

J. Vogel[1], K. Wackerbauer[2], and U. Stahl[1]

[1]Fachgebiet für Mikrobiologie und Genetik, Technische Universität Berlin, Gustav-Meyer-Allee 25, D–13355 Berlin, Germany
[2]Forschungsinstitut für Brauerei und Mälzerei der Versuchs- und Lehranstalt für Brauerei in Berlin, Seestrasse 13, D–13353 Berlin, Germany

During fermentation of beer, yeast cells produce ethanol as well as diacetyl, a substance with a low taste threshold of about 0.2 ppm (1). This unpleasant butter-like flavor can be removed by means of a separate maturation period of 2-6 weeks during which yeast cells degrade the diacetyl. Recombinant DNA techniques have made it possible to reduce the amount of α-acetolactate, the precursor molecule of diacetyl, in yeast. The α-acetolactatedecarboxylase (ALDC) gene from *Acetobacter pasteurianus*, which decarboxylates α-acetolactate directly into acetoin without forming diacetyl, was isolated and transferred into brewer`s yeast. Test fermentation with recombinant yeast having ALDC activity showed no differences in fermentation properties. At the end of the fermentation process hardly any diacetyl was measurable; thus the subsequent maturation period was unnecessary.

Microorganisms play a significant role in food biotechnology. They are used in brewing, baking, dairy products as well as for the production of desired flavors. Lactic acid bacteria are used to produce products such as yogurt, cottage cheese and butter milk. They are also used as starter cultures, not only for lactic acid production, but also for the production of diacetyl to give a "genuine" butter flavor.

Soy products as an alternative to meat have become more popular over the last few years. Anyone who has eaten such a "meat substitute" will testify that it tastes like meat. This is due to the *Rhizopus* and *Mucor* species which possess proteolytic activity and therefore degrade the soy protein into peptides, which then confers the final product a meat-like flavor.

Some fungi species are able to produce familiar, or pleasant odor, one example being a smell reminiscent of coconut which is often emitted when *Trichoderma* species have been grown on non-defined media. It is thus not surprising that efforts have been made to use microorganisms as producers of odor compounds for the food industry.

Lactones are widely used in the flavoring industry and are characterized by being generally pleasing in odor and flavor. They are described as being fruity, coconut-like, buttery, nut-like or sweet. A variety of yeast and filamentous fungi such as *Penicillium notatum* produce lactones when incubated with keto acids. To produce this odor or flavor by chemical means requires seven successive steps, so the advantage of using fungi to produce lactones is quite obvious. Another advantage of using biological flavors is, that biotransformation reactions are stereoselective so only one enantiomer is formed and no contaminations of other unpleasant substances are present. In addition, in Germany a "natural" flavor compound is more likely to be accepted by the food administration authorities and the public than a chemical one.

Some fungi produce flavor compounds which are unpleasant and not wanted in the final product. They are regarded as contamination and might even be toxic. It is therefore necessary to remove these undesired compounds. Technological improvements such as the optimizing of temperature or the supplementing of nutrients may be one method of solving the problem. However, as the metabolism is ultimately dependent on its genetic constitution, a genetic approach could also be a solution.

One possibility could be to interrupt the pathway in which an unpleasant flavor compound is synthesized by using mutagenesis methods. This would lead to the accumulation of an intermediate product which must be flavorless and non toxic for the cell in higher concentrations. Another possibility is the conversion of an unpleasant compound by the cells into a neutral substance, which must of course also be flavorless and nontoxic for the cell. For this approach a gene transfer is necessary, which is only possible by genetic engineering techniques.

Diacetyl, an Off-Flavor Compound Produced by Yeast During Fermentation of Beer

Yeast cells used in the brewing process affect the flavor produced. Although beer making is a long-established, traditional process, difficulties occasionally occur in obtaining a product of steady quality. One reason for this is that during fermentation yeast cells, apart from producing ethanol, also produce "off flavor compounds" such as dimethylsulfide (reminiscent of boiled vegetables), hydroxy-2-butenolide ("Maggi"-like flavor) and diacetyl which arises from α–acetolactate, an intermediate of the valine-isoleucine-pathway. In the main fermentation α–acetolactate leaks out of the yeast cells into the wort, where it is spontaneously oxidatively decarboxylated to diacetyl. Since this compound can be used by the cell as an electron-acceptor, it is taken up again by the yeast cells and reduced via acetoin to 2,3-butanediol. The non-enzymatic conversion of α–acetolactate to diacetyl is a slow reaction, so by the end of the main fermentation α–acetolactate remains in the wort and therefore a maturation must follow. During the maturation α–acetolactate is converted to diacetyl, which is then incorporated into the cells and reduced to acetoin (Figure 1). Because of the low temperature and reduced yeast cell numbers during the maturation the conversion of α–acetolactate to diacetyl and the incorporation of diacetyl into the cells takes time, but is necessary during the production of lager beer.

Diacetyl has a butter-like flavor with a low taste threshold of about 0.2 ppm. To remove this compound a maturation period of 2-6 weeks is required in order to allow the diacetyl to be reabsorbed into the yeast cells. This makes the brewing process costly.

Strategies to Reduce the Diacetyl Content in Beer

A way to reduce lagering time would be by simply adding purified bacterial ALDC to the fermenting wort. This enzyme which decarboxylates α–acetolactate directly to acetoin without forming diacetyl has been found in several gram-negative and gram-positive bacteria such as *Lactobacillus lactis, Acetobacter pasteurianus, Bacillus brevis* and *Enterobacter aerogenes (2)*. Beer fermented in the presence of ALDC purified from *Enterobacter aerogenes* showed no analytical differences to normal beer, except that the

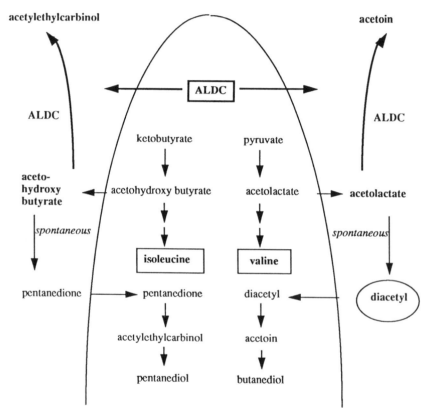

Figure 1: Pathway of the isoleucine- and valine-synthesis and the formation of viccinal diketones by yeast.

diacetyl concentration was very low and therefore lagering time could be reduced from some weeks to several hours (*3*). Although this procedure is advantageous, two points must be considered. Firstly, commercially prepared ALDC has to be added which makes the process more costly and secondly, the addition of any enzymes during the brewing process is prohibited in Germany and would contravene the German beer-brewing purity laws ("Deutsches Reinheitsgebot").

Since it is not possible to decrease the spontaneous conversion rate from α–acetolactate to diacetyl, one has to consider whether it may be possible to decrease the α–acetolactate pool in the yeast itself by strain improvement. With respect to the isoleucine-valine pathway, different strategies are possible which could reduce the α–acetolactate pool in yeast. One could decrease the activity of the acetolactate synthase so that less α–acetolactate is formed (*4*). The alternative is the molecular cloning of an additional ILV5 gene coding for reductoisomerase, into yeast, which results in an increased reductoisomerase activity (*5*). Both methods interfere with the amino acid pathway of valine, leucine and isoleucine and could have negative influences on growth.

Another strategy for reducing the α–acetolactate-pool in the yeast or the acetolactate-content in beer is the isolation and transfer of the ALDC-gene from an appropriate organism to yeast. In 1988 the ALDC-gene was firstly isolated from *Enterobacter aerogenes* (*6*), followed by a corresponding gene of *Klebsiella*(*7*). Both genes could be expressed in yeast and, as expected, the diacetyl-content was reduced. Since both, *Enterobacter* and *Klebsiella* belong to the Enterobacteriaceae, using a GRAS (general recommended as safe) organism would increase the chances of obtaining approval for the transformed yeast by authorities, especially since *Acetobacter pasteurianus,* a bacterium widely used in food industry to produce vinegar, could be chosen as the source of the ALDC-gene.

Cloning and Analysis of the Acetolactatedecarboxylase-Gene from *Acetobacter pasteurianus*

In order to clone the ALDC-gene the total DNA of *A. pasteurianus* was isolated, restricted with HindIII and randomly integrated into the LacZ-gene of the *E. coli* vector pUC19. After transfer into the *E. coli* recipient NM522, approximately 2000 colonies were assayed using the Voges-Proskauer reaction, for their ability to form acetoin which was not present in untransformed cells.

In analyzing one of the acetoin-positive clones it was found that it carried an *Acetobacter*-fragment of about 3 kb in size. As it can be assumed that only a part of the fragment is responsible for the formation of acetoin, it was sequentially truncated and found that an EcoRV/HindIII-fragment of only 1kb in length was necessary for ALDC expression. Sequencing experiments revealed a 879bp ORF on the EcoRV/HindIII fragment which seems to code for the ALDC (Figure 2).

Expression of the Acetolactatedecarboxylase-Gene in Laboratory Yeast Strains

In order to render the ALDC-gene from *Acetobacter pasteurianus* expressible in yeast it was integrated into an expression-cassette consisting of the alcoholdehydrogenase I(ADH1)-promoter and the tryptophan I(trp)-terminator. In addition, the plasmid contains the G418 resistance-gene for the selection of plasmid-carrying yeast cells and a 2μm origin which is necessary for autonomous replication in yeast (Figure 3).

As might be expected in transformed yeast cells, only a slight ALDC-activity could be detected because four start codons are located at the putative beginning of the ORF.

```
CCCCGGCCAG CCCCAGCCGG ATATGTCCCA GCTCACCCCG GGCCAGACCG TGGGCGGTCA  60
CCAGAAACTG GGCCTGCAGG TCCAGAATGG TGCGGGCGCG GGGCAGGAGG GTCGCGCCAA 120
TTTCGGTCAG CTTCACACCG CGCGTCTGCC GTTCAAACAG CGACGCCAGT TTTTGCTCCA 180
GCTGCTTGAT CTGCTGGCTG AGGGGCGGCT GTTCCATCCC TAGCTCTTCC GCAGCCCGGG 240
TAATACTGCC GTGGTCTGCC ACGGTGACAA AGTAACGCAG GTGCCTGATA TCCATGTGCC 300
                                                  EcoRV
ATATTTATAG GATATGGATA ATGCGTTCAA TATATATTGG AAGACTAGAT CCCGCTGTAA 360
TAACGTCTGT GTCATTGAGA TCAGCAACAA TGAGTTTGGA ACAGTGCGCA CGATGAAGAA 420
CAGCCCGGTG GCAGACATGG ACGTTCGGTC TTCTGCACTC GGAAACGGTG TGGGTAAGAA 480
ACCCGTCGCC AACCGTCTTT ATCAGACCTC CACCATGGCC GCTCTGCTGG ATGCCGTGTA 540
CGATGGCGAA ACCACGCTGG ACGAAGTGCT GCACCACGGC AATTTCGGCC TTGGCACGTT 600
TAACGCGCTG GATGGCGAAA TGATTGTGAC CGATGGTGTC GCACGCCAGT TCCGTGCGGA 660
AGGGCAGGCT GCCGAGGTTC CCGGTTCTCT CAAAACGCCT TTTGCCTGCG TGACATATTT 720
TGAGCCGGAA AAAACGCTCA ATATTGATAC ACCGCAGACA AAAGAAACAT TTGAAGCACT 780
GGTCGACCAG TTGGTGGGTA ATCCCAACCT GTTTGGTGCC GTTCGCTTTA CCGGGCAGTT 840
TGAGCGGGTG GATACGCGCA CGGTGTTCTG CCAGTGCAAG CCCTATCCGC ACATGCTGGA 900
TGTTGTGAAA AAGCAGCCCA CTCTGACCAT GGAATCCGTG ACCGGCACCA TGATCGGCTT 960
CCGCACCCCG GTTTATATGC AGGGTGTGAA CGTGGCGGGT TATCATCTGC ACTTCCTGAC 1020
GGAAGACCAG AAACGCGGTG GGCACGTGAC GGAATACCGG CTGGTGCGTG GCCAGCTTGA 1080
GGTTGCCGTG ATCTCCGATC TTGAAATTCA GCTGCCGCGC ACAGAGCAGT TTGCAAAAGC 1140
AAACCTTAAT CCTGAGCATC TGAGTGAAGC CATTCGGATT CGGCAAGGCG GCTGAAGCTT 1200
                                                            HindIII
```

Figure 2: The complete nucleotide sequence of the ALDC-gene and its 5′flanking sequence.

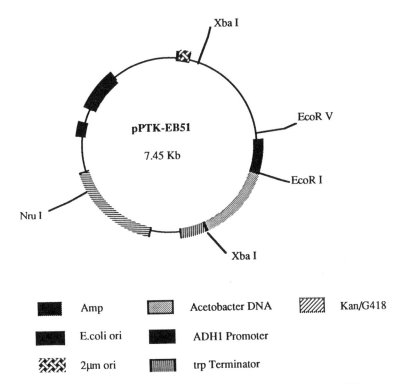

Figure 3: Physical map of the plasmid pPTK-EB51, containing a 1kb *Aceto-bacter pasteurianus* DNA fragment.

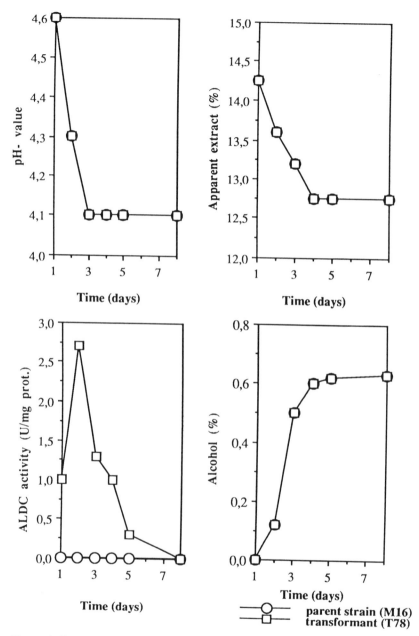

Figure 4: Fermentation characteristics of the transformed yeast strain (T78) and its parental strain (M16).

In order to try to increase the amount of ALDC transcripted, the first ATG (which is out of frame), the second codon (which is in frame with the fourth one) and the third ATG (which is in frame with the first one) were deleted. Whereas the absence of the first and the second codon had no effect on ALDC-activity, the additional deletion of the third increased the activity six-fold. It is therefore obvious that the last of the four ATG codons is the translation initiation codon of the ALDC-gene which is 780bp in length.

Fermentation Experiments with Laboratory Yeast Strains

In order to investigate the ALDC-activity during fermentation, the yeast strains containing the ALDC were compared with its parental strain in 500ml scale fermentations at 20°C for 8 days.

As shown in Figure 4, there was no substantial difference in fermentation performance. The pH value in both strains decreased from 4.6 in the first day to 4.1 by the fourth day of fermentation, and remained thereafter constant. The progress of the apparent extract was as expected - a decrease until the fourth day and then a constant level. Ethanol was produced by both strains in the same manner. The low ethanol concentration of 0.6 % at the end of fermentation was unexpected. This behavior seems to be characteristic for the lab-strain used and is of course useless for practical purposes. However, during main fermentation (until the fourth day) strong ALDC activity could be detected in comparison with the untransformed strain. The amount of ALDC was obviously sufficient for keeping the concentration of diacetyl and of α–acetolactate in the medium low. The amount of diacetyl is less than 0.01ppm in contrast to 0.13 ppm in untransformed cells. The combined values for diacetyl and α–acetolactate in the transformed and untransformed cells are 0.06 ppm and 0.34 ppm, respectively.

The stability of the replicative vector used throughout all experiments was not very high. Under non-selective anaerobic conditions about 40% of the population had lost the ALDC-gene within 5 days which is equivalent to about 3 to 5 generations. For practical purposes such a stability is too low.

These results revealed, that it is possible to express the ALDC-gene of *Acetobacter* in a laboratory yeast strain which was subsequently able to convert α–acetolactate to acetoin. Hence the diacetyl- and α–acetolactate-content decreases about 15% of the initial concentration.

Expression of the Acetolactatedecarboxylase-Gene in Brewer's Yeast

Transformation of brewer's yeast with the mentioned construct was carried out and, unfortunately, all the obtained transformants not only propagated very slowly, they also had to be cultivated in complete medium: they were all auxotrophic for valine, leucine and isoleucine. They were not able to sythesize sufficient amounts of these amino acids probably due to the lack of intracellular α–acetolactate. However, as brewer's yeast takes up these amino acids slowly the transformants show reduced growth.

Secretion of the Acetolactatedecarboxylase into the Medium

To convert only the excreted, and therefore surplus, α–acetolactate into acetoin, the ALDC-gene was integrated into an expression-secretion-cassette comprising the tryptophan I-terminator and alcoholdehydrogenase I-promoter. This promoter is linked to the secretion signal sequence of the mating pheromone α–factor encoding gene

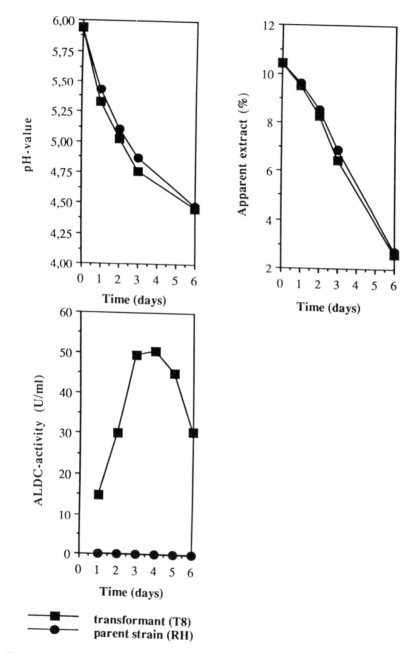

Figure 5: Fermentation characteristics of the transformed brewer´s yeast strain (T8) and its parental strain (RH).

(MFa1). In addition, the plasmid contains the G418 resistance gene and the 2μm origin. This approach should have no negative influence on amino acid synthesis. This construct worked well in laboratory yeast strains and, as expected, was also expressible in brewer's yeast. All obtained transformants propagate like the untransformed yeast cells and no auxotrophy appeared.

In addition, we were able to detect ALDC activity in the medium in which the transformants were cultivated, thus clearing obstacles for carrying out fermentation experiments with these transformants.

Fermentation Experiments with Brewer's Yeast

These experiments were carried out in EBC columns filled with 2.5 liters of yeast-containing wort at 10°C for 7 days .

As shown in Figure 5, there was no substantial differences in the fermentation performance. The pH value in both strains decreased from 5.9 in the first day to 4.4 by the sixth day of fermentation. They were also no differences in the progress of the apparent extract. During fermentation a strong ALDC-activity could be detected when compared to the untransformed strains. As may be seen, it increased until the fourth day of fermentation, however, at the end of the process activity decreased to a lower level. This is most probably due to pH instability of the enzyme. Other experiments showed that the enzyme is degraded at a pH value of about 4.6 and is constantly stable when potassium phosphate-buffered wort is used for fermentation.

However, the amount of the ALDC was sufficient for keeping the concentration of both diacetyl and α-acetolactate in the medium low. The amount of diacetyl was less than 0.01 versus 0.07 ppm in untransformed cells. The data for combined diacetyl and α-acetolactate in transformed and untransformed strains are 0.05 and 0.43 ppm, respectively. The analysis of beer fermented with the control and the transformant are shown in Table I. It is apparent that they hardly differ in characteristics such as alcohol content, color and bitter units, higher alcohols and esters, which are involved in creating the flavor characteristic of the final beer.

Table I. Analysis of the young beer fermented with transformed brewer's yeast strain (T8) and its parental strain (RH)

	diacetyl (actual) (ppm)	diacetyl (total) (ppm)	EtOH (vol%)	app. attenuation (%)	color (EBC)	bitter units (BE)	higher alcohols (ppm)	ester (ppm)
RH	0.07	0.43	4.28	76.9	16.5	23.0	87.7	29.0
T8	>0.01	0.05	4.33	77.6	16.5	22.4	88.3	30.3

Because brewer's yeasts are repeatedly used for beer production, ALDC activity has to be stably maintained. In order to observe the stability of the plasmid carrying yeast cells, three successive fermentations were carried out. Although the number of plasmid-carrying yeast cells decreases from 80% during the first fermentation to 69% at completion of the third fermentation, again only 0.06 ppm of total diacetyl was detectable versus 0.31 ppm in control cells.

Other experiments have shown that after 30 generations cultivated under non-selective conditions, 68% of the yeast cells still contained the plasmid. So, if, as mentioned earlier, during one fermentation 3 to 5 generations arise, at least 6 to 10 subsequent fermentation processes are possible, without any diacetyl flavor problems arising.

When using such a genetically modified yeast strain in the brewing process the diacetyl content should be much lower than the taste threshold after main fermentation, thus rendering subsequent maturation unnecessary.

Safety Assessment

Since the ALDC-gene is widely distributed among other *Aceto-* and *Lactic acid bacteria*, ALDC can be found in products such as vinegar, yogurt and butter - thus no risks arise in using such a gene for yeast strain improvement. In addition, when released into the environment ALDC-gene-carrying yeast cells have a selective disadvantage when compared to the wild type as they are handicapped in the regeneration of their reduction-equivalents (which are produced by the oxidation of the carbon source).

No risks are involved with regards to ecological aspects when using an antibiotic resistance gene as no evidence is available that natural transformation occurs among yeasts and so the direct transfer of genetic material into other organisms is not likely.

Transformation of this gene without using any markers, a process which we are at present using in our institute, makes it possible to obtain yeast strains carrying only sequences particular to the yeast and the ALDC-gene from an acetic acid-producing strain.

In conclusion, there are no health or environmental risks involved when using such a genetically improved yeast strain, so that production facilities originally designed for fermentation with non-recombinant yeast can be used.

Acknowledgments

We wish to thank Roslin Bensmann for secreterial assistance. This work was supported by AIF (Arbeitsgemeinschaft Industrieller Forschung).

Literature Cited

(1) Scherrer, A. *Schweizer Brauerei-Rundschau* **1972,** *82*, 1-23
(2) Godtfredsen, S.E.; Lorck, H.; Sigsgaard, P. *Carsberg Res. Commun.* **1983a,** *48*, 239-247
(3) Godtfredsen, S.E.; Ottesen, M.; Sigsgaard, P.; Erdal, K.; Mathiasen, T.; Ahrenst-Larsen, B. *EBC Congress* **1983b,** Lect. No. *17*, 161-168
(4) Ramos-Jeunehomme, C.; Masschelein, C.A. *Eur. Brew. Congr. Amsterdam* **1977,** 267-283
(5) Villanueba, K.D.; Goosens, E.; Masschelein, C. A. *ASBC J.* **1990,** *48*, 111-114
(6) Sone, H., T.; Kondo, K.; Shimizu, F.; Tanaka, J.; Inoue, T. *Techn. Rep. Kirin* **1988,** *31*, 11-18
(7) Blomquist, K.; Suihko, M.-L.; Knowles, J.; Penttilä, M. *Appl. Environ. Microbiol.* **1991,** *57*, 2796-2804

RECEIVED July 7, 1995

Chapter 15

Genetic Modification of Brewer's Yeast To Produce Acetolactate Decarboxylase and the Safety Aspects of the Beer Brewed by the Transformed Yeast

R. Takahashi[1], M. Kawasaki[2], H. Sone[1], and S. Yamano[1]

[1]Central Laboratories for Key Technology,
Kirin Brewery Co., Ltd., Fukuura, Kanazawa, Yokohama 236, Japan
[2]Research Center for Product Safety and Assessment,
Kirin Brewery Co., Ltd., Miyahara-cho, Takasaki 370–12, Japan

Brewer's yeast possessing bacterial acetolactate decarboxylase (ALDC) gene was constructed by genetic engineering. This yeast has the ability to convert acetolactate, the precursor of diacetyl, one of the most common off-flavors in beer, to acetoin that has no effect on beer flavor. The donor organism of the ALDC gene was *Acetobacter aceti* which is naturally found in vinegar production. The ALDC gene was integrated into the yeast chromosome and all undesired DNA sequences such as antibiotic resistance gene were removed from the transformed yeast. The introduction of the ALDC gene did not affect the characteristics of the transformed yeast with the exception of the ability to produce ALDC. Laboratory-scale fermentation tests showed that there was no significant difference between the beer produced from the transformed yeast and the beer from the parent yeast with the exception of diacetyl concentration.

The brewer's yeast currently in use is the result of selection over a 5000 year period of beer brewing. In the 1880s, Emil Christian Hansen isolated the first pure culture of brewer's yeast. It enabled brewers to start improving the yeast strains used for beer production. Most brewer's yeasts are prototrophic, homothallic, polyploid and have low sporulation ability and low spore viability, so it has been difficult to develop new strains of yeast by conventional genetic techniques such as mutation, mating, cytoduction and protoplast fusion. Even now, selection is the major method used to obtain a good brewer's yeast. However, the development of molecular genetics has made it possible to breed brewer's yeast by the application of recombinant DNA technology. An improved ability to ferment carbohydrates will make the utilization of raw materials more cost effective and make possible the production of low calorie beer. To achieve these goals, glucoamylase genes were introduced into brewer's yeast (1-3), and maltose utilization was improved by the introduction of a permease gene (4). Beer flavor can be improved by constructing a high acetate ester producing yeast (5). Yeasts producing low levels of the undesirable off-flavors, diacetyl (6-9) and hydrogen sulfide (10) have also been constructed. To improve the beer production process, the gene encoding the β-glucan degrading enzyme was introduced into brewer's yeast (11, 12). Yeast flocculation genes were isolated for the control of flocculation (13). While no recombinant brewer's yeast is used in production at present, such yeasts have been successfully tested through pilot scale fermentation and have demonstrated their usefulness.

In this report, we describe the breeding of a genetically modified yeast producing low diacetyl levels and discuss its safety.

The Aim of Introducing ALDC into Brewer's Yeast

Diacetyl is one of the most common off-flavors found in beer. At present, reduction in diacetyl concentration is mainly achieved by control of the fermentation temperature and the length of the beer maturation process. The beer fermentation process is divided into two stages; main fermentation and maturation. The fermentable sugars in wort are converted to ethanol during the main fermentation. The maturation process requires about 6 weeks. Most of the maturation processes occur within the first ten days. However, removal of diacetyl extends the maturation time to almost 6 weeks; therefore reduction of diacetyl is the rate limiting factor in beer maturation.

The formation and removal of diacetyl during beer fermentation are shown in Figure 1. The dotted arrows indicate the pathway from acetolactate to acetoin in traditional brewing. Diacetyl is formed from acetolactate, an intermediate of the isoleucine-valine pathway in yeast. Most acetolactate is converted into valine and leucine. A small fraction leaks into wort and is converted to diacetyl by nonenzymatic oxidative decarboxylation. During maturation, diacetyl is taken up by the yeast and reduced to acetoin by the enzyme diacetyl reductase. Acetoin has no effect on beer flavor. The conversion of acetolactate to diacetyl is the rate-limiting step. If the amount of acetolactate produced could be reduced, then we would obtain a reduction in the diacetyl concentration in the beer and a shorter maturation period.

The enzyme, acetolactate decarboxylase (ALDC), converts acetolactate directly to acetoin. It is found in various microorganisms, but not in yeast. Our aim was to introduce this enzyme into brewer's yeast to accelerate beer maturation by reducing the acetolactate concentration in young beer. We also wished to confirm that acetolactate production is the only process affected by genetic modification.

Construction of the ALDC Producing Brewer's Yeast

The transformed yeast was constructed to possess only the ALDC gene and DNA sequences necessary for expressing ALDC gene as introduced genetic materials. Antibiotic resistance marker gene, which is considered to be one of the greatest concerns on the safety of genetically modified organisms, was removed from the transformant. The ALDC expression cassette, which consists of the ALDC gene, promoter and terminator sequences, was integrated into yeast chromosome in order to produce ALDC stably. The site of integration was ribosomal DNA locus, because the ALDC expression cassette can be maintained stably for 150 generations at the locus (14). There are about 140 copies of the ribosomal DNA sequence in a yeast cell, so integration at this locus was considered to have no influence on the yeast physiology.

Procedure of genetic modification. The strategy to introduce the ALDC gene into brewer's yeast is shown in Figure 2. The ALDC gene was cloned from *Acetobacter aceti subsp. xylinum.* The gene was linked to the promoter and terminator of the *Saccharomyces cerevisiae* phosphoglycerate kinase gene for expression. Ribosomal DNA sequences were linked to the expression cassette for integration into the yeast genome by homologous recombination. The selection plasmid carries the G418 resistance gene and the replication origin in yeast (2μm DNA ori). First, brewer's yeast was co-transformed with the ALDC expression cassette fragment and the selection plasmid. Then the G418 resistant transformants obtained were screened to select those exhibiting ALDC activity. The ALDC expression cassette was integrated into ribosomal DNA sequences of yeast genome by homologous recombination. A co-transformed yeast strain was then cured of the selection plasmid by growth under non-selective conditions.

Laboratory-scale fermentation test. The effect of introducing ALDC was confirmed by a laboratory-scale fermentation test. The fermentation test was carried out at 8°C in 1 liter scale. The diacetyl concentration was significantly lower in wort fermented with the transformant than with the parent yeast (Figure 3). The maturation process of beer production was expected to be accelerated by using the transformed yeast.

Safety Aspects of the Genetically Modified Yeast

Safety of the genetically modified yeast was considered with several aspects. First, the safety of the host and donor organism was discussed. Second, the integration of the ALDC gene into yeast ribosomal DNA sequences and the removal of undesired DNA sequences were studied. If the transformed yeast was constructed to our specifications, safety consideration on antibiotic resistance marker gene would not be necessary and the properties of the transformed yeast would not change with the exception of ALDC production. Finally, laboratory-scale fermentation test was carried out in order to confirm it.

Host and donor organisms. The host, brewer's yeast, is a type of *Saccharomyces cerevisiae.* Not only is it used in commercial brewing but it is also used as a food source. It has a long history of safe use in human food consumption, with no reports of pathogenicity or toxin-producing activity. The donor organism, *Acetobacter aceti subsp. xylinum* was selected because it is naturally found in vinegar production. There is no report to suggest its pathogenicity or toxin-producing activity.

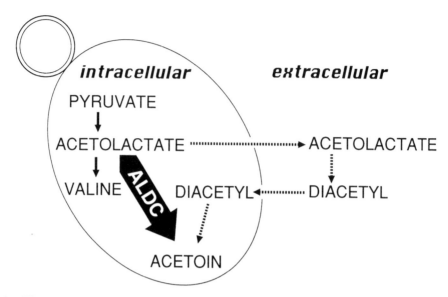

Figure 1. Formation and reduction pathway of diacetyl by brewer's yeast and proposed pathway for acetolactate decarboxylation in ALDC-producing yeast.

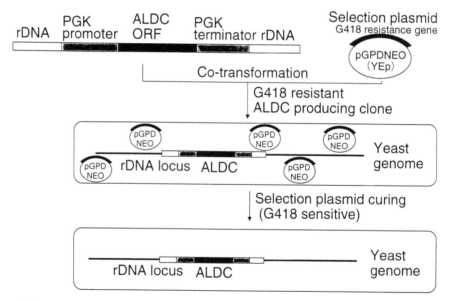

Figure 2. Introduction of ALDC expression cassette and curing of selection plasmid.

Introduced gene and the resulting gene product. The ALDC gene was cloned from *Acetobacter aceti subsp. xylinum*. The nucleotide sequences of the ALDC gene were determined. The promoter and terminator of the phosphoglycerate kinase gene were obtained from *Saccharomyces cerevisiae*. The ALDC gene product converts acetolactate directly to acetoin, which occurs naturally in beer. The location of the gene product expressed in the transformed yeast was intracellular.

DNA analysis of the transformant. DNA analysis was carried out in order to confirm that the yeast was genetically modified to our specifications. The location of ALDC gene and the removal of G418 resistance gene were examined by Southern analysis.

Integration of the ALDC expression cassette. Integration of the ALDC expression cassette into the ribosomal DNA locus of brewer's yeast was confirmed. First, the transformant's chromosomes were separated by pulse field gel electrophoresis. The integrated ALDC expression cassette was shown by Southern analysis to be integrated into chromosome XII, the same chromosome where the ribosomal DNA gene is located.
  The site of integration of the ALDC expression cassette was studied by restriction enzyme analysis using *Eco*RI, *Hin*dIII and *Bgl*II. The restriction enzyme fragment possessing the ALDC gene was detected using ALDC gene as a probe. Figure 4 indicates that the ALDC expression cassette was integrated into the ribosomal DNA gene as we expected.

Removal of undesirable DNA sequences. The removal of the selective plasmid containing the G418 resistance gene and *E. coli* derived sequences was confirmed by Southern analysis using G418 resistance gene as a probe. This indicated that the ALDC expression cassette was the only genetic material introduced into the yeast.

Characteristics of the transformant. The characteristics of the transformant were studied in order to confirm that ALDC production was the only difference between the transformed yeast and the parent yeast. The transformant showed the same characteristics as the parent strain in all tests carried out. There was no difference in the fermentation profiles of the transformant and parent strain (Figure 5a). Since ALDC converts acetolactate, an intermediate of the isoleucine and valine biosynthesis pathway to acetoin, amino acid consumption by the transformant was compared with that of the parent strain by measuring the amino acid concentration in the fermented wort. Valine and isoleucine consumption by the transformant were almost identical to the parent strain (Figure 5b). The flavor volatiles important for beer quality were also studied. No significant difference was observed between the parent and the transformant.
  When applying genetic engineering to brewer's yeast, it is important not to change any fermentation characteristics except for those characteristics which one aims to change. These results suggest that we successfully engineered a low-acetolactate producing brewer's yeast which fermented beer normally in laboratory scale fermentations.

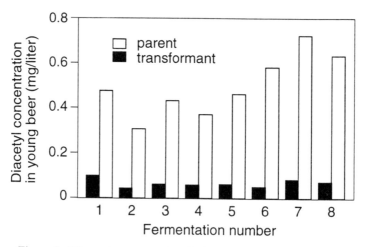

Figure 3. Diacetyl concentration during successive fermentations.

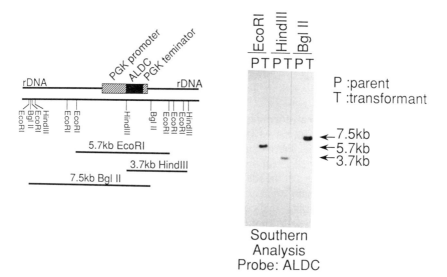

Southern
Analysis
Probe: ALDC

Figure 4. Restriction enzyme analysis of integrated ALDC expression cassette.

Genetic stability. Since brewer's yeast is used successively in practical beer production, 8 successive fermentations were carried out using the transformant. As shown in Figure 3, even after 8 fermentations, diacetyl concentration in young beer produced by the transformant was significantly lower than that produced by the parent strain. This means that the genetically modified character, the ability to produce ALDC, is stably maintained in the transformant.

Analysis of the Beer Brewed by the Transformed Yeast

As the ALDC expression cassette was maintained stably at the ribosomal DNA locus, the composition of the beer was expected to be equivalent except for diacetyl concentration. Some constituents of the beer were determined after laboratory-scale fermentation and maturation. The analysis includes basic beer components and some trace components considered to pose safety problems, because increased levels of natural toxicants is one of the most important subjects for the safety assessment of food produced by genetic modification.

The saccharides and organic acids composition of the beer. The saccharides and organic acids are important constituents in beer. These compounds are a good index for yeast activity and beer taste. Table I shows that there was no significant difference in the profile of monosaccharides, saccharides with more than four

Table I. Saccharides and Organic Acids in Beer

Component		Parent (ppm)	Transformant(ppm)
Saccharides	Arabinose	44	45
	Galactose	<20	<20
	Glucose	<20	<20
	Xylose	53	50
	Fructose	58	44
	Ribose	42	47
	Maltose	2200	1122
	Maltotriose	536	338
	Maltotetraose	1350	1334
	Maltopentaose	376	349
	Maltohexaose	333	325
	Maltoheptaose	238	241
Organic Acids	L-Lactic acid	60	69
	Acetic acid	253	219
	Pyruvic acid	74	105
	Tartaric acid	45	41
	Malic acid	93	77

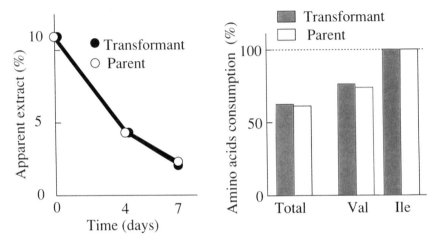

Figure 5.   Fermentation profile and amino acid consumption of the transformed yeast.

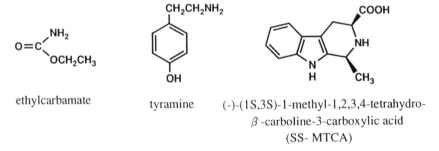

ethylcarbamate   tyramine   (-)-(1S,3S)-1-methyl-1,2,3,4-tetrahydro-
                                           β-carboline-3-carboxylic acid
                                           (SS- MTCA)

Figure 6. Natural toxicants in fermented foods.

Table II. Natural Toxicants in Beer

Compounds	Parent(ppm)	Transformant (ppm)
Ethyl carbamate	<0.001	<0.001
Tyramine	1.1	1.1
SS-MTCA[a]	0.44	0.41

[a](-)-(1S,3S)-1-methyl-1,2,3,4-tetrahydro-β-carboline-3-carboxylic acid

glucose units and organic acids. The concentration of maltose and maltotriose were slightly lower in the beer brewed with transformed yeast. These differences in concentration, however, are sometimes observed in laboratory-scale fermentations.

Natural Toxicants found in fermented foods. As beer is a fermented food, some toxic compounds naturally found in fermented foods were determined. The compounds analyzed are illustrated in Figure 6. Ethyl carbamate is a potential carcinogen naturally found at low ppb level in various fermented foods (15,16). Tyramine (17,18), a decarboxylation product of tyrosine, is thought to cause hypertension and is a mutagen precursor. SS-MTCA(19,20), a mutagen precursor, is the condensation product of tryptophan and acetaldehyde, both of which exist in most fermented foods.

   As shown in Table II, the levels of these compounds were remarkably low and did not increase in comparison with the beer produced by the parent yeast. These results demonstrate that there was no significant difference between the beer produced by the transformed yeast and the beer produced by the parent yeast with the exception of diacetyl concentration.

Conclusion

In applying recombinant DNA technology to the food product, no significant change except for the introduced trait is expected to occur. Transformed yeast was carefully constructed to possess no undesirable sequences and to show the same characteristics as parent yeast with the exception of ALDC production. The transformed yeast possessed only the ALDC gene and yeast DNA sequences necessary for expressing the ALDC gene as introduced genetic materials. The ALDC gene was stably maintained at the ribosomal DNA locus in the yeast chromosome. Laboratory scale fermentation tests showed that the total diacetyl concentration of the wort fermented by the transformed yeast was significantly lower than that by the parent yeast. The fermentation properties of the transformed yeast and the composition of some beer constituents did not change significantly with the exception of ALDC expression. These results demonstrate that there is no evidence that the introduction of ALDC brings about safety problems to the beer. Large scale fermentation and extensive analytical studies will ensure the safety of the beer brewed by the transformed yeast.

Literature Cited

1. Sakai, K.; Fukui, S.; Yabuuchi, S.; Aoyagi, S.; Tsumura, Y. *J. Am. Soc. Brew. Chem.* **1989**, *47*, 87-91.
2. Park, C.S.; Park, Y.J.; Lee, Y.H.; Park, K.J.; Kang, H.S.; Pek, U.H. *MBAA Technical Quarterly*, **1990**, *27*, 112- 116.
3. Yucum, R. R. *Proceeding of Biology Exposition 1986* , Butterworth, Stoneham, 171.
4. Shibano, Y.; Kodama, Y.; Fukui, N.; Nakatani, K. *BREWERS DIGEST-May*, **1993**, 23.
5. Fujii, T.; Bogaki, T.; Iwamatsu, A.; Minetoki, T.; Yashimoto, H.; Tamai, Y.; Nagasawa, N.; Hamachi, M.; Nunokawa, Y. *BREWERS DIGEST-May*, **1993**, 20.
6. Sone, H.; Fujii, T.; Kondo, K.; Shimizu, F.; Tanaka, J.; Inoue, T. *Appl. Environ. Microbiol.* **1988**, *54*, 38-42.
7. Yamano, S.; Tanaka, J.; Inoue, T. *J. Biotechnol.* **1994**, *32*, 173-178.
8. Penttilä, M.; Suihko, M.-L.; Blomqvist, K.; Nikkola, M.; Knowles J. K.C.; Enari.T.-M. *Yeast*, **1988**, Special Issue, 5473.
9. Blomqvist, K.; Suihko, M.,-L.; Knowles, J.; Penttilä, M. *Appl. Environ. Microbiol.* **1991**, *57*, 2797-2803.
10. Tezuka, H.; Mori, T.; Okamura, Y.; Kitabatake, K.; Tsumura, Y. *J. Am. Soc. Brew. Chem.* **1992**, *50*, 130- 133.
11. Suihko, M.-L.; Lehtinen, U.; Zurbriggen, B.; Vilpola, A.; Knowles, J.; Penttilä, M. *Appl. Microbiol. Biotechnol.* **1991**, *35*, 781-787.
12. Penttilä, M. E.; Suihko M.-L.; Lehtinen, U.; Nikkola, M.; Knowles J. K. C. *Curr. Genet.* **1987**, *12*, 413-420.
13. Watari, J.; Takata, Y.; Murakami, J.; Koshino, S. *Agric. Biol. Chem.* **1991**, *55*, 1547-1552.
14. Fujii, T.; Kondo, K.; Shimizu, F.; Sone, H.; Tanaka, J.; Inoue, T. *Appl. Environ. Microbiol.* **1990**, *56*, 997-1003.
15. Ough, C. S. *J. Agric. Food Chem.* **1976**, *24*, 323-328.
16. Dennis, M. J.; Howarth, N.; Key, P. E.; Pointer, M.; Massey, R. C. *Food Addit. Contam.* **1989**, *6*, 383-389.
17. Ochiai, M.; Wakabayashi, K.; Nagao, M.; Sugimura, T. *Gann*, **1984**, *75*, 1-3.
18. Cerutti, G.; Finole, C.; Vecchio, A.; Maccagnola, P. *Monatsschr. Brauwiss.* **1987**, *40*, 369-372.
19. Wakabayashi, K.; Ochiai, M.; Saito, H.; Tsuda, M.; Suwa, Y.; Nagao, M.; Sugimura, T. *Proc. Natl. Acad. Sci. U. S. A.* **1983**, *80*, 2912-2916.
20. Herraiz, T.; Ough, C.S. *J. Agric. Food Chem.* **1993**, *41*, 959-964.

RECEIVED June 7, 1995

Chapter 16

Safety Aspects of Genetically Modified Lactic Acid Bacteria

W. P. Hammes, C. Hertel, and C. Cavadini

Institute of Food Technology, Hohenheim University,
D–70593 Stuttgart, Germany

Lactic acid bacteria (LAB) are in several ways related to human life among which their contribution to virtually all processes of food fermentations is of special importance. The safe nature of the organisms involved is known from experience and genetic modifications have to keep this status unchanged. To improve the technological and beneficial potential of LAB, genetic modification has been applied to these organisms. The majority of studies was devoted to lactococci with application in the dairy field. With these organisms the development of food grade systems for genetic modification has become most advanced employing chromosomal gene integration and food grade vectors. Lactobacilli have broadest application in the various fields of food processing but comparatively little is known on their genetics. In studies with strains of *Lactobacillus sake* and *Lactobacillus curvatus* suitable as starters in meat fermentation, we studied the behavior of modified strains containing genes coding for useful properties such as formation of catalase and lysostaphin. The investigation of the stability of the constructs and of horizontal gene transfer was performed *in vitro* and in fermenting sausages. It was observed that the constructs retained their acquired metabolic activity and no gene transfer could be detected. On the other hand, the fermentation substrates permit efficient gene transfer as it was indicated by the transfer of the conjugative plasmid pAMß1.

For safety assessment of genetically modified organisms as hosts of newly acquired genes information is important concerning their systematic position, metabolic potential, genetics, habitats, pathogenicity, toxigenicity, and safe tradition of use. Research of lactic acid bacteria (LAB) has attracted increasing attention during the past 20 years, and the knowledge acquired has reached a status that not only permits to provide this information but also made the organisms widely applied in starter cultures and accessible to genetic modification. The practical use of the genetically modified LAB requires a careful assessment of their safety. It is the purpose of this communication to present an overview on the safety considerations of these organisms. As examples for experimental safety investigation we have selected results of our studies on lactobacilli.

The Relation of Lactic Acid Bacteria to Human Life

LAB are a group of evolutionary related organisms as it can be derived from the 16S rRNA similarity depicted in Figure 1 (*1*). They have in common a fermentative metabolism, forming from carbohydrates lactic acid as the major end-product. Minor metabolic variations are common and characterize strains or species with regard to their responses to ecological conditions, their potential to utilize substrates, and to form specific products others than lactic acid. These specific differences determine not only their adaptation to a substrate or habitat but also their specific effects in these environments. Thus, a metabolic diversity with a corresponding genetic background exists and gene technology intends to combine desired traits in strains suitable for technical or health promoting purposes of application.

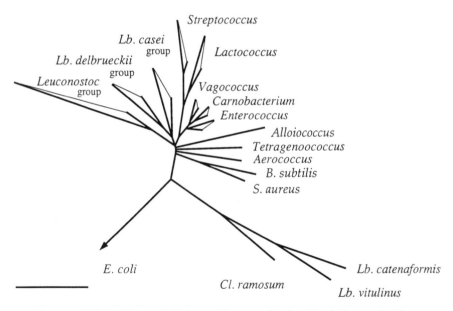

Figure 1. 16S rRNA based phylogenetic tree reflecting the phylogenetic relationships of lactic acid bacteria and selected relatives of low G+C content. The bar indicates 0.05 phylogenetic distances. Abbreviations: B, *Bacillus*; Cl, *Clostridium*; E, *Escherichia*; Lb, *Lactobacillus*; S, *Staphylococcus*. (Reproduced with permission from ref. 1)

The organisms of the group include species pathogenic for humans in the genus *Streptococcus*, and for animals in the genera *Streptococcus, Carnobacterium,* and *Aerococcus*. Furthermore, some strains of species belonging to the genera *Lactobacillus, Leuconostoc, Pediococcus, Lactococcus, Bifidobacterium, Aerococcus,* and *Enterococcus* have been isolated from patients which usually had a reduced host defense potential (*2, 3*). Some species are commensals of humans and exert either noxious effects on health (within the genus *Streptococcus*, e.g. causing dental plaques) or are health promoting living on mucosa and in the intestines (within *Lactobacillus* and *Bifidobacterium*) (*4, 5*). Finally, there is an important group of species that interferes with food and feed. In these man-made habitats they may either cause spoilage or perform the desired fermentations. Remarkably, no true food poisoning organisms have been found within the LAB, except for a potential of some species to form biogenic amines (*6*) and, furthermore, no reports on LAB-related allergies have been published.

Lactic Acid Bacteria in Food Fermentation

LAB participate in virtually all fermentation processes and contribute to the acquisition of properties of foods shown in Table I.

Table I. Effects of LAB on the properties of fermented foods (*4, 5*)

Desired effect of LAB	Food
sensory quality	
- flavor	all fermented foods
- texture	cheese, yoghurt, fermented sausages, and vegetables
- colour	fermented sausages
shelflife	all fermented foods
hygienic safety	all fermented foods
gas formation	sourdough, some cheeses
nutritional quality	fermented milk, vegetables, bread
"processing aid"	sour wort, sourdough

The food products and organisms involved in the fermentative process are compiled in Table II. From this table it can be derived that characteristic species are involved in the fermentation of the various substrates in which they are especially competitive and suitable to achieve the desired effects shown in Table I.

Starter cultures have been developed to control the fermentation processes and their use is characteristic for modern food-biotechnology. These cultures are commercially available and of practical importance for the fermentative processing of milk, meat, dough, wine and, to a minor extent, vegetables. A genetic modification of lactic acid bacteria used as starters aims at achieving the purposes shown in Table III.

Table II. Foods of plant and animal origin fermented by lactic acid bacteria either alone or in combination with other groups of microorganisms

Substrate	Product	LAB involved [a]
Plant origin		
Olives	Fermented olives	*Leuconostoc mesenteroides*
Cabbage	Sauerkraut	*Lactobacillus bavaricus*
Cucumber	Pickled cucumber	*Lactobacillus brevis*
Tomatoes and minor important substrates, e.g. celery, egg-plants, levant garlic, green beans	Fermented vegetables	*Lactobacillus curvatus* *Lactobacillus plantarum* *Lactobacillus sake* *Pediococcus pentosaceus*
Doughs or batters made from flour	Sourdough	*Lactobacillus sanfrancisco* *Lactobacillus brevis* *Lactobacillus pontis* sp. nov.
	Kisra sourdough	*Lactobacillus reuteri* *Lactobacillus fermentum* *Lactobacillus amylovorus*
Must or wine	Malolactic fermented wine	*Leuconostoc oenos*
Soy	Soy sauce	*Tetragenococcus halophilus*
Animal origin		
Milk	Cultured buttermilk	*Leuconostoc mesenteroides* subsp. *cremoris* *Lactococcus lactis* subsp. *cremoris* *L. lactis* subsp. *diacetylactis*
	Yoghurt	*Lactobacillus delbrueckii* subsp. *bulgaricus* *Streptococcus thermophilus*
	Cheese	*Lactococcus lactis* subsp. *lactis* *L. lactis* subsp. *diacetylactis* *Leuconostoc mesenteroides* subsp. *cremoris* *Lactobacillus helveticus* *L. delbrueckii* subsp. *lactis* *Lactobacillus casei* *Lactobacillus plantarum*
Meat	Fermented sausages	*Lactobacillus curvatus* *Lactobacillus plantarum* *Lactobacillus sake* *Pediococcus acidilactici* *Pediococcus pentosaceus*
Fish	Fish sauce	*Tetragenococcus halophilus*

Source: Reprinted with permission from ref. 5. Copyright 1994.
[a] In the course of the fermentation process these species occur at varying ratios

Table III. Aims of genetic modification of lactic acid bacteria used in food production

1. **Reduction of hygienic risks**
   e.g. formation of antagonistic compounds to inhibit growth of food poisoning organisms, removal of toxins from raw materials or of microbial origin
2. **Improvement of the nutritional value**
   e.g. foods enriched in vitamins or amino acids
3. **Exploitation of probiotic effects**
   e.g. by overcoming lactose intolerance, stabilization of the intestinal flora, exploitation of antitumor activity
4. **Performance of processes under improved ecological aspects**
   e.g. saving of energy, exploitation of new resources including substances presently considered as waste
5. **Improving process safety**
   e.g. by phage resistance, decreasing the risk of loss of metabolic properties by integration of plasmid encoded genes into the chromosome, improving the competitiveness of the starter organisms
6. **Simplification of microbial fermentative events**
   e.g. by combination of various properties in one single organism, e.g. combination of the activities of catalase, nitrate and nitrite reductase, lipase, proteinase in lactic acid bacteria
7. **Improvement of the ecological adaptation**
   e.g. by production of bacteriocins
8. **Reduction of costs**
   e.g. shortened process time, reduced input of energy
9. **Improvement of the efficiency of the cultures**
   e.g. increasing or acquisition of desirable properties such as enlarged metabolic spectrum, prototrophy, stronger or new aroma, color stability
10. **Access to new products**
    e.g. preparations of fermented fruit or vegetable juices, metabolites and enzymes.

Safety Considerations on the Use of Genetically Modified Lactic Acid Bacteria

Foreseeable potential risks that may arise from the genetic modification of LAB involved in food fermentation include acquisition of pathogenicity, allergenicity, toxigenicity, and creation of environmental hazards. These potential risks can be considered as very low. For example, LAB of importance in food fermentations are no pathogens. However, some strains of defined species (*vide supra*) have been isolated from patients (*2, 3*). Similarly, neither allergenicity nor toxigenicity are known for food LAB, except for a potential to form biogenic amines which is found in some species (*6*). Finally, as LAB are reduced in their metabolic activities and require growth factors and rich substrates, a spread of the organisms causing environmental hazards is rather unlikely. To exclude potential risks arising from the application of gene technology on LAB, the application of the strategy of using food-grade systems has been recommended (7). The elements characterizing this system are compiled in Table IV.

The Host of the Foreign Gene. The host is an organism with a tradition of safe use in food fermentations or as a non-noxious commensal of man. In view of the more general considerations described above, there is sufficient knowledge to select species and strains that can be generally recognized as safe. It is important that the genetic manipulation does not cause a loss in essential properties of the organisms

Table IV. Characteristics of food-grade systems

---

1) The Host
    A.   a safe organism
    B.   expresses the marker gene in cases where it is required
    C.   no loss of the technologically essential properties after the genetic modification

2) The food-grade marker gene
    A.   well defined (known sequence, regulation and stability)
    B.   endows the host with a selective advantage
    C.   acceptable in foods (not coding for resistance against chemotherapeutics)
    D.   originates preferentially from the host species (self-cloning) or from other GRAS organisms

3) The foreign gene
    A.   well defined (known sequence, regulation and stability)
    B.   endows the host with a new trait (see Table III)
    C.   originates preferentially from the host species (self-cloning) or from other GRAS organisms

4) The vectors
    A.   well defined (known sequence, copy number, stability, mobility, not conjugative)
    B.   derived preferentially from the host species (self-cloning) or from other
       GRAS organisms

5) The integration of the genes
    i) **Extrachromosomal integration of the foreign gene**
       using food-grade vectors
    ii) **Chromosomal integration of the foreign gene**
       integration by a single cross-over recombination event (e.g. Campbell-like mechanism, see Figure 2) employing food-grade integration vectors (8)
       integration of the foreign gene by a double cross-over recombination event (gene replacement, see Figure 2)

---

such as competitiveness, and contribution to maintaining a high hygienic standard of the products. In cases where the application of marker genes is necessary to maintain stability of an employed vector, it is essential that their activity is stably expressed under the fermentation conditions, when it is required.

The Food-grade Marker Gene. Food-grade markers may be essential in constructing the genetically modified strain and endow the host with a selective advantage. This advantage may also be extended to the application of the organism in food fermentations, for example, when the food is characterized by containing a unique substrate such as lactose in milk and the wild-type host does not utilize lactose. The food-grade marker gene should be well defined with regard to the sequence, mechanism of regulation, and its stable locational and segregational retainment. It should be further acceptable in foods and should, for example, not code for resistance against chemotherapeutics. It appears to be of special advantage to use marker genes, preferentially from host-species or at least from other GRAS organisms.

Other Elements. The characterization of foreign genes in a food-grade system resembles that of food-grade marker genes. The genes may be incorporated into the host either integrated in vectors or the chromosome. When a food-grade vector is employed, it should be well defined with regard to sequence, copy number, stability, mobility, and it should not be conjugative. Again, the elements of the vector should preferentially be derived from the host-species or at least from other GRAS organisms. As shown in Table V, food-grade vectors have already been constructed for LAB and may be used for practical applications.

Table V. Food-grade vectors developed for lactic acid bacteria

Vector	Marker gene	Host organism	Reference
pFM011	nsr coding for the nisin resistance determinant from *Lactococcus lactis*	*Lactococcus lactis*	9
pNZ305	lacF coding for Enzyme III^{lac} (PTS) from *Lactococcus lactis*	*Lactococcus lactis* mutated in lacF	7, 10
pPR602	thyA coding for thymidylate synthase from *Lactococcus lactis*	*Lactococcus lactis* mutated in thyA	11
pBG10	ß-galactosidase gene from *Lactobacillus bulgaricus*	*Lactobacillus helveticus*	12

As schematically depicted in Figure 2, a chromosomal integration of the foreign gene is possible by a single cross-over event, in which a food-grade integration vector is employed (8). An example for a so-called Campbell-type integration in LAB was provided by Leenhouts et al. (13). The authors transferred the proteinase genes prtP and prtM of *Lactococcus lactis* subsp. *cremoris* to a proteinase deficient strain of *L. lactis* subsp. *lactis* using a non-replicating integration vector. They could show that the genes were stably maintained in the new host and that the integrated plasmid was amplified up to 8-fold leading to an 11-fold increased proteolytic activity when compared with the activity of a strain in which the genes were located on a plasmid. When gene integration is based on this strategy, the chromosomal insert includes more information than required for a mere expression of the foreign gene.   In contrast, gene replacement recombination permits to incorporate into the chromosome nothing but the foreign gene, as it is schematically depicted in Figure 2. The mechanism involves a primary integration of the vector by excision of redundant information after a second cross-over event. The applicability of gene replacement recombination in LAB was demonstated with *Lactobacillus* helveticus *(14)*, *L. plantarum (15)*, *Lactococcus lactis (16, 17, 18)* and *Streptococcus thermophilus (19)*.

It is conceivable that the stability of maintenance of the foreign gene within the host is especially high when integrated into the chromosome. On the other hand, its localization in an expression vector may lead to segregational instability or even to its transfer to other species, when the nature of the vector permits conjugational transfer. This latter condition may be considered as worst case with regard to safety. It may be used as a tool to investigate gene transfer within the food environment in which the following factors may influence the frequency of conjugal transfer:

i)   The environment:
         For example, it may permit a close cell to cell contact which is a prerequisite for a plasmid transfer. In fact, in a liquid such as yoghurt no gene transfer was observed (see Table VI). On the other hand, on the surface of or in crevices between particles, as present in solid foods (e.g. sausage), the frequency can be rather high (see Table VI).
ii)   The numbers and the ratio of donor and receptor cells.
iii)   Relatedness of the strains.
iv)   Characteristics of the conjugative plasmid (e.g. the host range of the origin).
Experimental evidence for a transfer of conjugational plasmids in fermenting foods was obtained in investigations compiled in Table VI. The results obtained confirm the role of the above factors.

Table VI. Vertical transfer of conjugative plasmids of lactic acid bacteria in food

Food process and organism	Plasmid	Range of transfer frequencies*	Reference
cheese making *Lactococcus lactis*	pIL205	7.6×10^{-11} to 7.0×10^{-8}	20
sausage fermentation *Lactobacillus curvatus*	pAMß1	5.0×10^{-7} to 1.3×10^{-6}	21
yoghurt fermentation *Streptococcus thermophilus*	pAMß1	not detected	22

*, number of transconjugants per recipient cell at the time of sampling

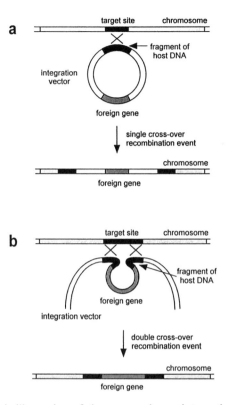

Figure 2. Schematic illustration of chromosomal gene integration. a) Campbell-type integration of non-replicating plasmids into the chromosome of the host by a single cross-over recombination event. b) Replacement integration of non-replicating plasmids into the chromosome of the host by a double cross-over recombination event. (Adapted from ref. 8)

In our study of conjugal transfer in fermenting sausages (*21*), we were interested to obtain information on the stability of plasmid maintenance, frequency of plasmid transfer in the food substrate and the performance of the strains under the aspect of practical usefulness of the modified starters. We had employed a plasmid cured strain of *Lactobacillus curvatus* endowed with one of four marker plasmids, one being plasmid pAMß1, a broad host range conjugative plasmid. The size of the inocula was chosen to permit growth for 6, 12 or 18 generations. As shown in Figure 3, the various strains grew well and, as indicated by the pH decrease in the sausages, achieved the essential technological effects. Two strains harboring plasmids pIL253 and pCA44, respectively, exhibited growth by one order of magnitude below that one of all other strains. These strains were also restricted in their capability to retain the acquired plasmids. This was revealed by analysis of the plasmid patterns of the isolates obtained from sausages investigated after 5 days of ripening as it is indicated by a shadowed area in Figure 3.

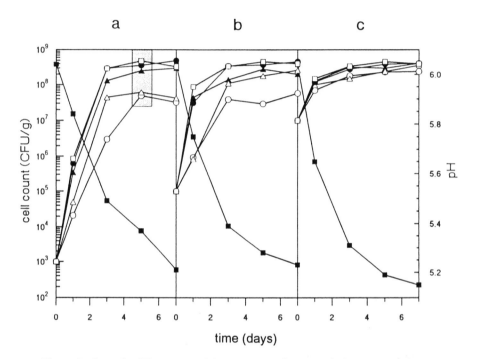

Figure 3. Growth of *L. curvatus* LTH1432 transformants during sausage fermentation. Cell count of *L. curvatus* LTH1432 containing (□) no plasmid, (●)
pAMß1, (▲) pNZ12, (Δ) pIL253 or (○) pCA44. (■) Decrease in pH. Shaded
area: analysis of these samples with respect to the percentage of plasmid containing, plasmid free and meat borne strains is shown as an example in Table
VII. The sausage mixtures were inoculated with fig a 10^3, fig b 10^5 and fig c 10^7
cfu/g of *L. curvatus* harbouring no for one of the indicated plasmids.
(Reproduced with permission from ref. 21. Copyright 1992)

Table VII. Segregational stability of a plasmids in *L. curvatus* during sausage fermentation[a] The percentage of strains is presented which either harbour marker plasmids conferring resistance to antibiotic or might have lost this property or have grown in the fermenting substrate as meat borne LAB.

Marker plasmid	Total cell count (cfu/g) in MRS	Percentage of antibiotic resistant	Percentage of plasmid containing[b]	Percentage of plasmid free	Percentage of meat borne cells
pAMß1	3.6×10^8	> 98	> 98	n.d.	n.d.
pNZ12	2.6×10^8	95	95	5	n.d.
pIL253	8.2×10^7	70	70	15	15
pCA44	1.0×10^8	50	50	25	25

Source: Reprinted with permission from ref. 21. Copyright 1992.
[a] Sausages were inoculated with 10^3cfu/g in 4 independent batches of sausage fementations employing the same raw materials. Samples were taken at the 5th day of fermentation, which corresponds to 18 generations (see also shaded area in Figure 3). The 100% value corresponds to the total count on MRS medium.
[b] Percentage of strains isolated from MRS without antibiotic which contained the marker plasmid. Meat borne strains contained different plasmids.
n.d. = not detected; >98 = virtually all cells

As shown in Table VII, a clear loss of plasmids had taken place in these strains and similar results were obtained in *in vitro* studies in which the strains were cultured for a time equivalent to hundred generations. We also performed co-inoculation experiments using equal cell counts of the *L. curvatus* strains harbouring plasmids pAMß1 and the relatively well maintained plasmid pNZ12, respectively. When inocula were employed of 5×10^2, 5×10^4 or 5×10^6 cfu/g, the strains were perfectly active in the fermenting sausages. As shown in Table VIII, plasmid pAMß1 was transferred to strains harbouring plasmid pNZ12. The frequencies of transfer at the time sampling depended on the size of the inoculum and were highest at use of 5×10^6 cfu/g. It is remarkable that the frequencies observed in sausages were in the same order of magnitude as those determined in *in vitro* studies under the optimum conditions of filter mating.

The investigation of a potential gene transfer was extended to studies in which expression vectors were employed. The vector pJK356 was constructed by Klein et al.(23) and was based on a cryptic plasmid present in the starter strain *L. curvatus* Lc2. We inserted into this plasmid (Hammes et al., Hohenheim University, unpublished results) the gene coding for lysostaphin originating from *Staphylococcus simulans* (24). Three constructs, pLS100, pLS200, and pLS201 were obtained upon transformation of *L. curvatus* which differed by the size of the gene insert. We could show that cultures containing transformed cells of *L. curvatus* (pLS100) actively killed staphylococci during the process of sausage fermentation and also in mayonnaise based salads. The cultures exhibit, therefore, a remarkable potential to prevent staphylococcal food poisoning. Furthermore, the catalase gene *katA* from *L. sake* (25) was cloned and successfully expressed in *L. curvatus* using the vector

Table VIII. Conjugation frequencies for pAMß1 between *Lactobacillus curvatus* LTH1432 (pAMß1) and *L. curvatus* (pNZ12)

Procedure	Conjugation frequency per recipient
filter mating	8×10^{-6}
broth mating 4°C	none
broth mating 30°C	5.5×10^{-7}
in fermenting sausage	
5×10^{6} cfu/g[a]	1.3×10^{-6} [b,c]
5×10^{4} cfu/g[a]	5×10^{-7} [b,d]
5×10^{2} cfu/g[a]	none

Source: Reprinted with permission from ref. 21. Copyright 1992.
[a] Cell count of recipient at the time of inoculation.
[b] The conjugation frequencies were calculated as number of transconjugants per recipient cells at the time samples were taken. They represent the average of the values determined at days [c]1, 3, 5 and 7 or [d]5 and 7, respectively. There were no significant differences in frequencies calculated for different days.
[d] No transconjugants were detected after the 1st and 3rd day of the fermentation.

pJK356 resulting in the plasmid pLSC300 (Hammes et al., Hohenheim University, unpublished results). The catalase activity is a desirable property for starter cultures to minimize deleterious effects of hydrogen peroxide.

The segregational stability of the vector pJK356 in *L. curvatus* was investigated by continuous transfer of cells in broth culture. The vector without insert (pJK356) is retained rather stably for nearly 100 generations. On the other hand, if the vector carries an insert, the plasmid (pLS100 or pLSC300) is gradually lost. No structural instabilities of the vector and the newly constructed plasmids were observed. A loss of an introduced plasmid during food fermentation is not necessarily of disadvantage at practical application. It appears to be sufficient that the plasmids are harbored as long as they are needed in the process and this comprises usually not more than 20 generations.

We have finally investigated if the vectors pJK356 and pLSC300 can be transferred from *L. curvatus* to another strain of *Lactobacillus*. As a positive control for a transfer potential, we employed *L. curvatus* strains harboring the conjugative plasmid pAMß1. The results obtained are compiled in Table IX. Again, the frequencies for transfer of plasmid pAMß1 are similar in filter mating and sausage fermentation. Neither a conjugal transfer nor pAMß1-associated mobilization of the vectors could be detected.

Conclusion

The genetic modification of LAB is possible, as it was revealed from the study of various species. The modified strains can contribute to improve food quality and process technology. Strategies for a construction of strains have been proposed which are suitable to provide safe strains.

Table IX. Transfer frequencies of plasmids pJK356, pLSC300 and pAMß1 from *Lactobacillus curvatus* Lc2c (pJK356, pAMß1) or *L. curvatus* Lc2c (pLSC300, pAMß1) to *L. curvatus* Lc2c (Rif ^{R}, Str ^{R})

Procedure	Transfer frequency per recipient with		
	pJK356	pAMß1	pLSC300
filter mating	none	5.3×10^{-6} ($< 1 \times 10^{-8}$)	n.i.*
during sausage fermentation (inoculation rate: 1 x 10^7 cfu/g)	none	7.5×10^{-6} ($<7.9 \times 10^{-8}$)	
during sausage fermentation (inoculation rate: 1 x 10^7 cfu/g)		5.2×10^{-6}	none ($<4.4 \times 10^{-8}$)

* n.i., not investigated;

Acknowledgments. This work was supported by grants from the Bundesministerium für Forschung und Technologie (0319280).

Literature Cited

1. Schleifer, K.H.; Ludwig, W. In *The lactic acid bacteria Vol. 2; The genera of lactic acid bacteria*; Wood, B.J.B.; Holzapfel, W.H., Eds.; Blackie Academic & Professional (an imprint of Chapman & Hall): London, New York, **1995** (in press).
2. Gasser, F. *Bull. Inst. Pasteur.* **1994**, *92*, 45-67.
3. Aguirre, M.; Collins, M.D. *J. Appl. Bact.* **1993**, *75*, 95-107.
4. Hammes, W.P.; Weis, N.; Holzapfel, W. In *The Prokaryotes*; Balows, A.; Trüper, H.G.; Dworkin, M.; Harder, W.; Schleifer, K.H., Eds.; 2nd edition; Springer: New York, **1992**, pp 1535-1594.
5. Hammes, W.P.; Tichaczek, P.S. *Z. Lebensm. Unters. Forsch.* **1994**, *198*, 193-201.
6. Straub, W.; Kicherer, M.; Schilcher, S.M.; Hammes, W.P. *Z. Lebensm. Unters. Forsch.*, in press.
7. De Vos, W.M.; Simons, G.F.M. In *Genetics and Biotechnology of Lactic Acid Bacteria*; Gasson, M.J.; de Vos, W.M., Eds.; Blackie Academic & Professional (an imprint of Chapman & Hall): London, New York, **1994**, pp 52-105.
8. Venema, G. *J. Dairy Sci.* **1993**, *76*, 2133-2144.
9. Froseth, B.R.; McKay, L.L. *J. Dairy Sci.* **1991**, *74*, 1445-1453.

10.   De Vos, W.M.; Boerrigter, I.; van Rooijen, R.J.; Reiche, B.; Hengstenberg, W. *J. Biol. Chem.* **1990**, *265*, 22554-22560.
11.   Ross, P.; O'Gara; Condon, S. *Appl. Environ. Microbiol.* **1990**, *56*, 2164-2169.
12.   Hashiba, H.; Takiguchi, R.; Jyoho, K.; Aoyama, K. *Biosci. Biotech. Biochem.* **1992**, *56*, 190-194.
13.   Leenhouts, K.J.; Gietema, J.; Kok, J.; Venema, G. *Appl. Environ. Microbiol.* **1991**, *57*, 2568-2575.
14.   Bhowmik, T.; Fernández, L.; Steele, J.L. *J. Bacteriol.* **1993**, *175*, 6341-6344.
15.   Hols, P.; Ferain, T.; Garmyn, D.; Bernard, N.; Delcour, J. *Appl. Environ. Microbiol.* **1994**, *60*, 1401-1413.
16.   Leenhouts, K.J.; Kok, J.; Venema, G. *J. Bacteriol.* **1991**, *173*, 4794-4798.
17.   Simons, G.; Nijhuis, M.; de Vos, W.M. *J. Bacteriol.* **1993**, *175*, 5168-5175.
18.   Biswas, I.; Gruss, A.; Ehrlich, S.D.; Maguin, E. *J. Bacteriol.* **1993**, *175*, 3628-3635.
19.   Mollet, B.; Knol, J.; Poolman, B.; Marciset, O.; Delley, M. *J. Bacteriol.* **1993**, *175*, 4315-4324.
20.   Gabin-Gauthier, K.; Gratadoux, J.J.; Richard, J. *FEMS Microbiol. Ecol.* **1991**, *85*, 133-140.
21.   Vogel, R.F.; Becke-Schmid, M.; Entgens, P.; Gaier, W.; Hammes, W.P. *System. Appl. Microbiol.* **1992**, *15*, 129-136.
22.   Kleinschmidt, J.; Soeding, B.; Teubner, M.; Neve, H. *System. Appl. Micobiol.* **1993**, *16*, 287-295.
23.   Klein, J.R.; Ulrich, C.; Plapp, R. *Plasmid* **1993**, *30*, 14-29.
24.   Heinrich, P.; Rosenstein, R.; Böhmer, M.; Sonner, M.; Götz, F. *Mol. Gen. Genet.* **1987**, *209*, 563-569.
25.   Knauf, H.J.; Vogel, R.F.; Hammes, W.P. *Appl. Environ. Microbiol.* **1992**, *58*, 832-839.

RECEIVED June 7, 1995

ENZYMES

Chapter 17

Enzymes from Genetically Modified Microorganisms

Sven Pedersen, Bent F. Jensen, and Steen T. Jørgensen

Novo Nordisk A/S, Novo Allé, DK–2880 Bagsvaerd, Denmark

Enzymes from genetically modified microorganisms play an increasing role in food technology. An example of such an enzyme with application possibilities both within the flavor and food area is cyclomaltodextrin glycosyltransferase (CGTase). A novel CGTase has been isolated from a strain of *Thermoanaerobacter*, a thermophilic anaerobe. The enzyme is extremely heat stable and has a temperature optimum of 90-95°C at pH 6.0. The gene encoding *Thermoanaerobacter sp.* CGTase has been transferred to a *Bacillus* host thus making possible large-scale production of the enzyme in commercially acceptable yields. This enzyme produces a mixture of α-, β-, and γ-cyclodextrins, not ideal from an industrial point of view, because β-CD's find an increasing industrial use. Possibilities of rationally designing mutants of CGTase, which mainly produce β-CD will be discussed. Safety aspects of the production of CGTase will be described.

1. Introduction

Food enzymes produced by genetically modified microorganisms have been used commercially for a number of years. Examples are a maltogenic amylase for production of maltose syrups (1), a lipase from *Mucor miehei* for interesterification of fats (2), and acetolactate decarboxylase (ALDC) for maturation of beer (3). The properties and safety aspects of a novel CGTase from *Thermoanaerobacter*, an enzyme that has been introduced for production of cyclodextrins for technical applications and which will be introduced during 1995 as a food enzyme, will be described in this chapter.

The necessary safety evaluations have been carried out for this enzyme, and the results are now being reported. A GRAS-petition will be filed for the enzyme in 1995, claiming that the CGTase, when produced according to current Good Manufacturing Practice, should be generally regarded as safe for use in the production of

0097–6156/95/0605–0196$12.00/0

cyclodextrins. To illustrate the safety evaluation of the CGTase, the safety tests carried out for another enzyme, a maltogenic amylase (4), will be described. The safety tests carried out for the two enzymes, the maltogenic amylase and the CGTase, are the same. The possibilities of improving the properties of the enzymes by protein engineering will be discussed.

2. Cyclodextrins

CGTase stands for cyclodextrin glycosyltransferase, describing the enzyme's primary function, to produce cyclodextrins from starch. Cyclodextrins are oligosaccharides with a closed ring structure, where the glucose units are joined together by α-1,4 linkages. Cyclodextrins containing 6, 7 or 8 glucose units are most common and are known as α, β and γ-cyclodextrins, respectively (Figure 1). The orientation of the molecule is such that the hydroxyl groups are on the outside of the ring structure and the interior of the cavity contains the C-H-groups and the glycosidic oxygens. Thus the cavity is hydrophobic while the external surfaces are hydrophilic. The dimensions of the ring are dependent on the number of glucose units (Figure 1). The hydrophobic interior is utilized in one of the main applications of cyclodextrins: flavors and spices form inclusion complexes with cyclodextrins and are protected against oxidation. Several spices, e.g., horseradish, are commercially available in Japan in this form. In addition to the production of cyclodextrins, CGTase catalyzes the transfer of glycosyl residues from a donor such as starch to a suitable acceptor. This is utilized commercially in glycosylation of the intense sweetener, stevioside (Figure 2). Stevioside is isolated from the leaves of the plant *Stevia rebaudiana*. The product is, however, bitter and its solubility is low. Glycosylation decreases bitterness and increases solubility. Examples of commercial applications of cyclodextrins are given in Table I.

Table I. Applications of cyclodextrins

Applications of cyclodextrins
Stabilization of volatile substances
- flavors, spices etc.
Modification of physical properties
- improve solubility of pharmaceuticals, eg., prostaglandin
- reduce bitterness
- mask unpleasant odors
Selective adsorption
- removal of cholesterol from egg, butter

3. Development of the *Thermoanaerobacter* CGTase

Processing of starch at concentrations of industrial interest requires jet cooking as an initial step (5). Jet cooking is a continuous processing step, where the starch with an added amylase is liquefied in steam jet cookers or similar equipment operating at temperatures up to 105-110 °C. When a conventional α-amylase is used for liquefaction the reaction products are maltodextrins, which can act as acceptors in the cyclization reaction, thus reducing the cyclodextrin yield (6). Liquefaction with the CGTase itself became possible in 1985, when a CGTase from the thermophilic anaerobic genus, *Ther-*

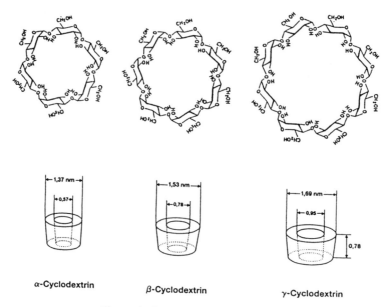

Figure 1. Structures of cyclodextrins

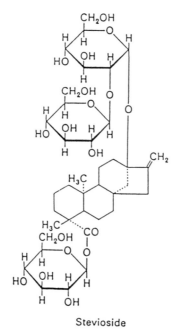

Stevioside

Figure 2. Stevioside

moanaerobacter was isolated (Figure 3)(7,8). This enzyme is active and stable at high temperatures and low pH, and there are no traces of low molecular weight oligosaccharides produced in the initial stages of the reaction. The *Thermoanaerobacter* enzyme has its temperature optimum at about 90°C (Figure 4) and a broad pH-curve with a maximum at about pH 5.8 (Figure 5). A modification of the Pharmacia Phadebas α-amylase method has been chosen as analytical method, because it is both rapid and simple. Starch hydrolysis is accompanied by the release of a blue colour, which is measured spectrophotometrically at 620 nm.

Anaerobic bacteria are poor enzyme producers and in order to produce the enzyme on an industrial scale gene transfer to a more suitable organism was chosen (7,8). The use of well known safe microorganisms with a long record of use in the food industry as host for cloning food enzymes may also be advantageous from a safety evaluation point of view.

The gene coding for the CGTase was cloned into *Escherichia coli* as an intermediate host, and then into *Bacillus subtilis*. Figure 6 shows a simplified diagram of the basic steps. In *Bacillus subtilis*, CGTase is expressed extracellularly in quantities reaching approximately 40-fold higher than in the original *Thermoanaerobacter* strain (7,8). Characterization of the recombinant CGTase relative to the native CGTase with respect to molecular weight (SDS-PAGE), isoelectric point, thermostability, action pattern, liquefaction activity, and cyclodextrin production indicated no differences between the enzymes. The recombinant CGTase cross-reacted with antibody raised against the native CGTase. After the initial cloning and expression in *B. subtilis*, the gene was transferred into a *Bacillus* strain suitable for large scale production. The DNA-sequence of the CGTase gene, which has been transferred to *B. subtilis* and later to the *Bacillus* production strain has been determined. The DNA sequencing has been made on DNA extracted from *E. coli*. The N-terminal aminoacid-sequence of the native enzyme purified from *Thermoanaerobacter* has been determined. The cloned and sequenced DNA codes for this N-terminal aminoacid-sequence.

4. Safety aspects
Manufacture and quality controls
Microbial food enzymes are produced by pure culture fermentation of carefully selected strains of microorganisms grown on steam sterilized natural substances. Fermentation conditions and conditions during recovery of the enzymes should be carefully controlled to ensure Good Manufacturing Practice (GMP) throughout the course of enzyme production. The final products must meet the food grade quality criteria formulated by the FAO/WHO Joint Expert Committee on Food Additives (JECFA)(9) and Food Chemicals Codex (FCC)(10). These criteria are summarized in Table II, and include evaluation of raw materials, additives and processing aids, determination of chemical contaminants, microbial contamination, mycotoxins in fungal enzymes and antibiotic activity in the final product.

Safety evaluation principles
The safety in use of microbial food enzymes is often based on a combination of a history of safe use and a safety evaluation based upon scientific procedures. The scientific procedures often include both in vivo and in vitro toxicology studies or chemical analysis for suspected toxic compounds in enzyme preparations (11).

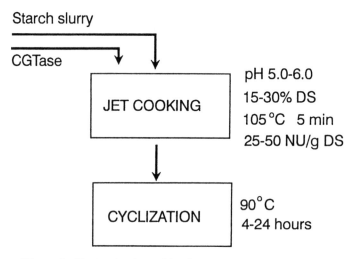

Figure 3. CD-Production with *Thermoanaerobacter* CGTase

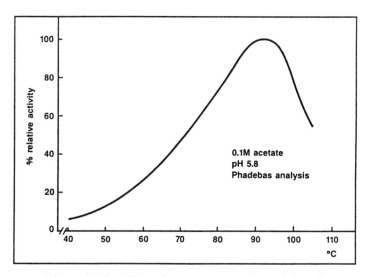

Figure 4. The effect of temperature on the activity of
Thermoanaerobacter CGTase

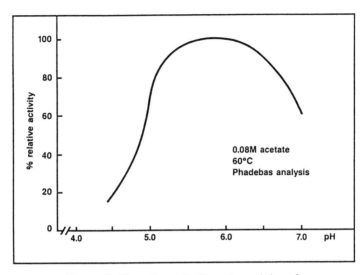

Figure 5. The effect of pH on the activity of
Thermoanaerobacter CGTase

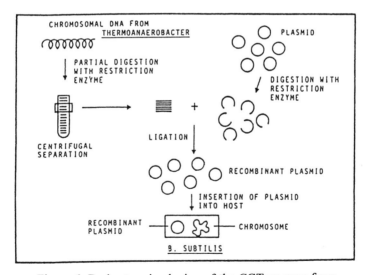

Figure 6. Basic steps in cloning of the CGTase gene from
Thermoanaerobacter into *Bacillus*

Table II. JEFCA and FCC specifications for enzymes

Parameter	JECFA	FCC
Raw materials	Free of harmful and undesirable substances	Same as JECFA
Additives and processing aids	Acceptable for food or insoluble and removed after use	Same as JECFA
Chemical contaminants	Lead $\leq$10 ppm Arsenic $\leq$ 3 ppm Heavy metals $\leq$40 ppm	Same as JECFA (ditto) (ditto)
Mycotoxins in fungal enzymes	Aflatoxin, Ochratoxin A Sterigmatocystin, T-2 toxin, Zearalenone must be absent	Ensure that products do not contain mycotoxins
Antibiotic activity	Negative	Not specified
Microbial contamination	Coliforms $\leq$30/g Salmonella negative E.coli negative Total viable count $\leq$50000/g	Same as JECFA (ditto) Not specified May cause no increase in food over that accepted for food

All enzymes currently used for food processing are per se safe for consumption, i.e., the enzyme protein in its active or inactivated form cannot be perceived to exert any toxiological reactions in man. Assuming that Good Manufacturing Practices are being followed during fermentation and recovery, toxic contaminants can then only come from the enzyme source organism itself. JECFA (12) classifies microbial food enzymes in 3 classes according to the type of source organism, and the tradition of use. The following toxicology programs are recommended (Table III).

Table III. JEFCA classification of microbial food enzymes

Class A : Enzymes derived from microorganisms that are traditionally accepted as constituents of foods or are normally used in the preparation of foods. These products are regarded as foods and, consequently, considered acceptable, provided satisfactory chemical and microbiological specifications can be established.
Class B : Enzymes derived from non-pathogenic microorganisms found as contaminants of foods. These materials are not considered as foods. It is necessary to establish chemical and microbiological specifications and to conduct short-term toxicity studies to ensure the absence of toxicity. Each preparation must be evaluated individually and an Acceptable Daily Intake (ADI) must be established.
Class C : Enzymes derived from microorganisms that are less well known. These materials also require chemical and microbial specifications and more extensive toxicological studies, including a long-term study in rodent species.

Enzymes produced industrially often belong to class B. In the case of food enzymes from genetically modified microorganisms the following components should be evaluated: the donor organism, the host organism, and the resulting cloned organism.

The donor organism
The strain should be precisely characterized in order to establish its taxonomic classification. The scientific literature is reviewed for reports of pathogenic potential or tests for suspected toxins or antibiotic substances.

The host organism
The host organisms used for cloning of industrial enzymes should be chosen from well known and widely used microorganisms in recombinant research. Sporulation-deficient derivatives of such strains, e.g., *Bacillus subtilis*, are thus exempt from the NIH-guidelines for Research Involving Recombinant DNA-molecules and can be considered safe host organisms for use in cloning (4).

The resulting cloned organism
The DNA insert from the donor organism should be thoroughly characterized. Information on the process of extra DNA, genetic characteristics such as markers, presence of dormant genes, genetic stability, gene transfer and antibiotic resistances should be used in the prediction of any undesirable effects on human health and ecological behaviour. The vector should be characterized at the DNA level and genetically with respect to genes found on the vector.

Safety studies
In 1982 the U. S. Food and Drug Administration (FDA) developed a system for determination of the minimum toxicology programs required for the safety assessment of food additives, the so-called "Red Book" (13). Applied to microbial food enzymes this system gives advice on the required toxicology program depending solely on the human exposure to the particular enzyme material. According to the European Union guidelines (14), the following tests are normally required for enzymes derived from microorganisms (Table IV).

Table IV. Tests normally required for enzymes from microorganisms

Tests normally required for enzymes from microorganisms:
(a) 90-day oral toxicity test in a rodent species;
(b) Two short-term tests:
1. A test for gene-mutations in bacteria
2. A test for chromosomal aberrations
(preferably *in vitro*).

The test material should be produced according to the procedure used to prepare the commercial preparations. To illustrate the safety tests carried out for food enzymes the tests carried out for maltogenic amylase are shown in Table V. The tests 'production of antibiotics, acute toxicity, dietary toxicity and mutagenic evaluation' are carried out for evaluation of the safe use, while the tests 'irritation and sensitization' are carried out for the evaluation of the safe handling of the enzymes. As already mentioned the safety tests carried out for the CGTase are quite similar. The tests are described in detail in (4).

Table V. Safety tests carried out for maltogenic amylase

Safety aspect	Type of test
Sensitization	Delayed contact hypersensitivity in guinea pigs
Production of antibiotics	Determination of antibacterial activity
Acute toxicity	Acute inhalation toxicity (4h exposure)
Dietary toxicity studies in rodents	Subacute toxicity (4-week dose and target organ finding study). Subchronic toxicity (90 days)
Mutagenic evaluation	Gene mutation in vitro Chromosome aberration in vivo
Irritation	Skin irritation (4h exposure) in rabbits Eye irritation in rabbits

The results from this study led to the conclusion that the maltogenic amylase, when produced according to current Good Manufacturing Practice should be considered generally recognized as safe for use in the production of maltose syrup. Similar conclusions can be drawn from the safety tests of the CGTase. A GRAS-petition containing the data will be filed in 1995 for this enzyme.

5. Protein Engineering of CGTase

The DNA sequence encoding a given enzyme may be specifically changed at various positions, to produce modified enzyme products that may have improved properties. This is termed protein engineering. From an industrial point of view the *Thermoanaerobacter* CGTase has the disadvantage,that it produces a mixture of α-, β-, and γ-cyclodextrins. In this respect it is similar to CGTases from other sources, although the major CD produced varies (Table VI).

Table VI. Some properties of CGTase from different sources

Source	Temperature optimum (°C)	pH optimum	Major CD
B. macerans	55	5.5	α
B. circulans	55	6.0	β
B. stearother- mophilus	70	6.0	α/β
B. subtilis	65	8.0	γ

The *Thermoanaerobacter* CGTase produces an approximately equal mixture of α- and β-cyclodextrin together with a small portion of γ-cyclodextrin (Figure 7).
Industrially two different approaches are used to produce pure cyclodextrins: crystallization of β-CD and complexation with organic solvents.

Crystallization of β-CD

This approach utilizes the differences in solubility of the cyclodextrins (Table VII). After the cyclization process the CGTase is inactivated (Figure 8), residual starch is hydrolyzed using glucoamylase and the β-CD crystallized. The products are pure β-CD and a mixture of glucose and all three types of CD's, the amounts depending on the specific CGTase being used.

Complexation with organic solvents

Precipitation of the cyclodextrins during reaction by complexation with water immiscible solvents shifts the equilibrium in favor of CD-production. Reasonably high yields of the specific cyclodextrins can be obtained this way. Known complexants for α-CD are for example 1-decanol, for β-CD for example toluene or cyclohexane (15) and for γ-CD the use of cyclododecanone has recently been reported (16). An example of production of α-CD using 1-decanol as complexing agent is shown in Figure 9.

Table VII. Properties of cyclodextrins

	α-CD	β-CD	γ-CD
Number of glucose units	6	7	8
Molecular weight	973	1135	1297
Cavity depth, Å	7.9-8.0	7.9-8.0	7.9-8.0
Cavity diameter (internal), Å	4.7-5.2	6.0-6.4	7.5-8.3
Water solubility, 25 °C, g/100 ml	14.5	1.85	23.2

It is only β-CD that is produced industrially by the use of organic solvents. The known complexants for α-CD have high boiling points (1-decanol 229°C), which make them difficult to remove from aqueous systems by evaporation and the complexants for γ-CD are too expensive for commercial use. Disadvantages of the solvent process are the toxicity of the solvents, the flammability and the need for a solvent recovery process. Because there is a great demand for a process that also could produce α- and γ-CD at economic prices and because the processes used for production of β-CD are not ideal, research is now being directed towards modification of the CGTase by protein engineering in order to develop mutants that are specific for one type of CD.

The 3D-structure of a CGTase enzyme from *Bacillus circulans* has recently been determined. The structure consists of five domains, where the A, B, and C-subdomains are similar in structure to the α-amylases (17,18,19). The largest is the A-domain with $(\beta\alpha)_8$-topology. Domain B is an extended loop between the third β-strand and third α-helix of the A domain and domain C has β-sheet structure. In addition to the A, B, and C-domains CGTase possesses two C-terminal domains, D and E, with β-sheet structure. The active site contains three acidic residues Asp 229, Glu 257 and Asp 328 in addition to Tyr 195 (18, 19, 20). The detailed knowledge of the 3-dimensional structure has made it possible to begin the rational design of mutants with improved product specificity (21, 22). The long term goal is to design 3 mutants producing only α-, β- and γ-CD, respectively. The commercial introduction of such a protein engineered enzyme is clearly at least a couple of years ahead.

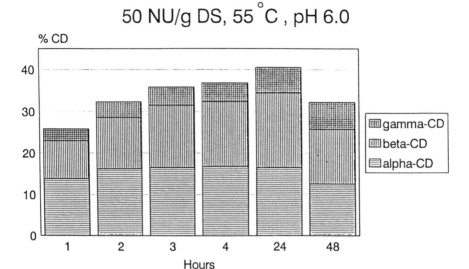

Figure 7. Production of α-, β-, and γ-cyclodextrin by
Thermoanaerobacter CGTase

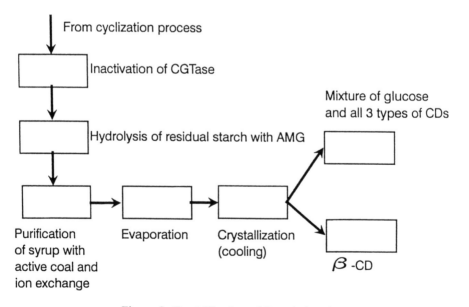

Figure 8. Crystallization of β-cyclodextrin

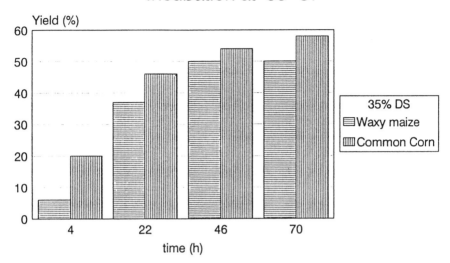

Figure 9. Yields of α-CD using 1-Decanol as complexation agent

6. Conclusions

The development of a commercial enzyme preparation has been described. The use of genetically modified microorganisms to produce this enzyme on an industrial level is necessary, because the enzyme was discovered in an anaerobic bacterium. It is not likely that an enzyme with the required high temperature stability can be found in microorganisms that are suitable for use as a source of food enzymes. The enzyme produces a mixture of α-, β-, and γ-CD, but industrially cyclodextrins are needed in pure form. It is possible to produce cyclodextrins in pure form by using organic solvents as complexing agents, but except for the production of β-CD these processes are expensive. The obvious alternative is to try to modify the enzyme in a way that it produces only one type of cyclodextrin. The detailed knowledge of the 3-dimensional structure has made it possible to begin the rational design of mutants with improved specificity, even though the commercial introduction is clearly several years ahead.

The safety evaluation of these enzymes from genetically modified micro-organisms has been illustrated with maltogenic amylase as an example. The evaluation includes the donor and host organism and careful assessment of the resulting cloned organism. Appropriate safety tests are carried out including 90-days oral toxicity test in a rodent species and two short term tests for gene mutation in bacteria and a test for chromosomal aberrations. The regulatory status of the *Thermoanaerobacter* CGTase is that the appropriate safety tests have been performed and the results are now being reported. It is expected that a GRAS-petition will be filed for the enzyme in 1995, claiming that the CGTase, when produced according to Good Manufacturing Practice, can be regarded as safe for use in the production of cyclodextrins.

Acknowledgements:
The figures are reproduced by permission of Novo Nordisk A/S, Denmark.

References:
1. Outtrup, H.; Normann, B.E. *Starch/Stärke* **1984**, 36, 405-411.
2. Boel, E.; Huge-Jensen, B.; Christensen, M.; Thim, L.; Fiil, N.P. *Lipids* **1988**, 23(7), 701-706.
3. Diderichsen, B.; Wedsted, U.; Hedegaard, L.; Jensen, B.R.; Sjøholm, C. *J. Bacteriology* **1990**, 172(8), 4315-4321.
4. Andersen, J.A.; Diderichsen, B. K.; Hjortkjaer, R. K.; De Boer, A. S.; Bootman, J,; West, H.; Ashby, R. *Journal of Food Protection* **1987**, 50(6), 521-526.
5. Sicard, P.J.; Saniez, M.-H. In *Cyclodextrins and their Industrial Uses*, Duchêne, D., Ed., Editions de Santé, Paris, **1987**, pp. 77-103.
6. Norman, B. E.; Jørgensen, S.T. *Denpun Kagaku* **1992**, 39(2), 101-108.
7. Starnes, R.L.; Flint, V.M.; Kalkocin, D.M. In *Minutes from the Fifth International Symposium on Cyclodextrins,* Duchêne, D., Ed., Editions de Santé, Paris, **1990**, pp. 55-61.
8. Starnes, R.L.; Amemiya, K.; Jørgensen, P.L.; Jørgensen, S. In *Minutes from the Sixth International Cyclodextrin Symposium*, Hedges, A.R., Ed., April 21-24, **1992**, Chicago.
9. JECFA, General specifications for enzyme preparations used in food processing. *FAO Food and Nutrition Paper* **1984**, 31/2, 129-131.
10. *Food Chemicals Codex,* National Academy of Science/National Research Council, Food and Nutrition Board, Committee on Codex Specifications, National Academy Press, Washington D.C., **1981**, pp. 107-110.
11. Jensen, B.F.; Eigtved, P. *Food Biotechnology* **1990**, 4(2), 699-725.
12. JECFA. *Principles for the Safety Assessment of Food Additives and Contaminants in Food*, Environmental Health Organisation, Geneva, **1987**, pp. 135-137.
13. FDA "Red Book", *Toxicological Principles for the Safety Assessment of Direct Food Additives and Color Additives Used in Food* , U.S. Food and Drug Administration, Bureau of Foods, **1982**.
14. Scientific Committee for Food, *Guidelines for the presentation on food enzymes,* Reports of the Scientific Committee for Food, **1991**, 27 series pp. 13-22.
15. Armbruster, F. C. In *Proceedings of the Fourth International Symposium on Cyclodextrins*, Szejtli, H. and J., Eds., Kluver Academic Publishers, Dordrect, **1988**, p. 33-39.
16. Rendleman, J.A.,Jr. *Carbohydrate Research*, **1993**, 247, 223-237.
17. Svensson, B.; Søgaard, M. *Biochemical Society Transactions* **1992**, 20, 34-42.
18. Klein, C.; Schultz, G. E. *J. Mol. Biol.* **1991**, 217: 737-750.
19. Klein, C.; Hollender, J.; Bender, H.; Georg E. Schultz. G.E. *Biochemistry* **1992**, 31, 8740-8746.
20. Lawson, C.L.; van Montfort, R.; Strokopytov, B.; Rozeboom, H. J.; Kalk, K.H.; de Vries, G.E.; Penninga, D.; Dijkhuizen, L.; Dijkstra, B.W. *J. Mol. Biol.* **1994**, 236, 590-600.
21. Nakamura, A.; Haga, K.; Ogawa, S.; Kuwano, K.; Kimura, K.; Yamane, K. *FEBS Letters* **1992**, 296(1): 37-40.
22. Fujiwara, S.; Kakihara, H.; Sakaguchi, K.; Imanaka, T. *Journal of Bacteriology* **1992**, 174(22): 7478-7481.

RECEIVED June 7, 1995

ANIMALS

Chapter 18

Progress in Genetic Modifications of Farm Animals

V. G. Pursel

Gene Evaluation and Mapping Laboratory, Livestock and Poultry Sciences Institute, Beltsville Agricultural Research Center, Agricultural Research Service, U.S. Department of Agriculture, Beltsville, MD 20705

For the past decade it has been possible to modify the genetic composition of farm animals by transfer of individual genes. Microinjection is the predominant method used to transfer genes into farm animals. Current research effort is devoted to improvement of productivity traits, enhancement of animal health, and production of biomedically useful human health products. Transgenic research initially involved genes coding for growth hormone (GH) and growth hormone releasing factor (GRF). More recent investigations have attempted to stimulate muscle development, to use bacterial enzymes so animals can synthesize certain essential amino acids, to induce expression of specific immunoglobulin or disease-resistance genes, and to direct expression of human proteins to the mammary gland, specific organs, or specific cells for production of useful human health products. The main limitations to progress are the lack of useful cloned genes for productivity traits and disease resistance and the insufficient knowledge of mechanisms involved in regulation of transgenes. Products produced by genetically modified animals will receive extensive scrutiny by regulatory agencies before consumption by humans is permitted. Currently no products are under evaluation for approval, but clinical evaluation of some rare biomedical products may begin soon.

The genetic composition of domestic animals has been manipulated for centuries to enhance their usefulness to humans. Development of recombinant DNA technology in the past decade has enabled scientists to isolate single genes, analyze and modify their nucleotide structures, make copies of these isolated genes, and transfer copies into the genome. An animal that integrates recombinant DNA in its genome is called "transgenic", and the transferred gene is called a "transgene".

Medically important human proteins have recently been produced in milk of transgenic sheep and pigs. Such products will soon begin clinical evaluation. Use of transgenic animals for food and fiber remains further in the future, because few agriculturally useful genes have thus far been isolated, sequenced, and cloned. In addition, greater knowledge of gene regulation is needed before some structural genes can be used to full benefit.

The purpose of this report will be to review the progress that has been achieved since transgenic modification of farm animals was first reported in 1985 (*1*), with emphasis on developments during the past few years.

Methods of Transferring Genes

Microinjection of Pronuclei. The primary method used to produce transgenic farm mammals is direct microinjection of DNA into the pronuclei of zygotes. Pronuclei of rabbit, sheep and goat zygotes can be readily seen using differential interference contrast (DIC) microscopy. Lipid granules in the cytoplasm interfere with visualization of pronuclei of pig and cow ova. Centrifugation of ova stratifies the cytoplasm so that pronuclei are visible with use of DIC microscopy (*2*).

To permit microinjection, ova are placed on a depression slide in a microdrop of media that is overlaid with silicone or paraffin oil to prevent evaporation. The microscope must be equipped with two micromanipulators, one for an egg-holding pipette and the other for an injection pipette. The holding pipette and injection pipette are each fitted with a tube leading to a syringe that permits either gentle suction or carefully controlled fluid injection. As an ovum is held with light suction by the holding pipette, the tip of the injection pipette is inserted through the zona pellucida and cytoplasm into the most visible pronucleus. Several hundred copies of the gene are expelled into the pronucleus. The person performing the injection carefully observes the pronucleus and withdraws the pipette when the pronuclear structure has visibly enlarged. Only one pronucleus is injected in pronuclear ova while both nuclei are injected in two-cell ova.

After microinjection, cow eggs are usually cultured in vitro until they are morulae or blastocysts before non-surgical transfer into the uterus of a synchronous host cow. The injected eggs of the other species are usually cultured only a few hours before they are transferred directly into the oviduct of synchronous host females. Pig ova can be transferred into the oviducts of donor females immediately after ova have been recovered without compromising the pregnancy rate or subsequent litter size (*3*).

The mechanism by which injected DNA integrates into a chromosome is unknown. Injected DNA usually integrates in a single site on a chromosome but multiple integrations can occur (*4,5*). Frequently, injected DNA results in multiple copies of the gene integrating in head-to-tail array. Recent research results indicate that integration rarely occurs during the first few cleavages (*6*). Breeding studies with transgenic pigs and sheep indicate mosaicism is a definite problem, with about 20% of founder transgenics failing to transmit the gene to progeny and another 20 to 30% transmitting the transgene to less than 50% of their progeny, presumably due to mosaicism in the germ cells (*7*).

The efficiency is usually lower for integration of transgenes into farm animals than into mice. The percentage of gene-injected embryos that develop into transgenic animals varied from 0.31 to 4.03% for pigs, 0.1 to 4.45% for sheep, 1.0 to 1.7% for goats, and 0.34 to 2.63% for cattle. The overall efficiencies are remarkably similar for farm animals (Table I). Transgenic chickens have been recently produced by microinjection of genes into the germinal disk of the recently fertilized egg (8,9). After microinjection, the chick embryo were cultured in a host eggshell until hatching time.

Table I. Efficiency of Producing Transgenic Farm Animals

Species	Experiments (No.)	Eggs Injected and Transferred (No.)	Percentage of Transferred Eggs		
			Born	Transgenic	Expressing
Cow	10	1,519[a]	16.8[a]	.86[a]	N/A
Goat	2	498	17.3	1.4	N/A
Sheep	10	5,242	10.6	.88	46.3
Pig	20	19,397	9.9	.91	52.3

[a] Transferred at morula and blastocyst stage.
N/A = not available.

Retroviral Insertion.   Retroviruses can be modified by recombinant DNA techniques to replace part of the viral DNA with a desired gene and then subsequently used as a gene vector. Embryos from several species have been successfully infected by retroviral vectors (10,11), but only transgenic mice and chickens have thus far been reported. Retroviral infection receives considerable attention because it offers several advantages over microinjection in certain applications. Principal advantages are: 1) integration of single copies of the gene without rearrangement at the site of integration; and 2) retroviral DNA integrates into a high percentage of embryos that can be infected by exposure to high concentrations of viral stock, by co-culture with infected cells in vitro, or, in the case of chickens, by microinjection into the blastodisk. The disadvantages are: 1) added work to produce a retrovirus carrying the transgene; 2) the gene being transferred must be smaller than 10 kb in size; 3) resulting transgenic animals are highly mosaic, which necessitates extensive outbreeding to establish pure transgenic lines; and   4) unresolved problems with expression of the transgene (12).

Stem Cell Insertion.   The third method of introducing genes into the germ line involves transfer of a gene into embryonic stem (ES) cells in culture and then incorporating these transgenic ES cells into an embryo, which is then chimeric for the transgene. The advantage of this procedure is that a particular genotype can be selected in vitro before introduction of the ES cells into the embryo. In addition, this technique is the only one that provides the ability for site-specific insertion of a transgene by homologous recombination (13).

A major disadvantage of this procedure at present is that many of the chimeric transgenic mice do not contain the transgene in their germ cells. Thus far, only transgenic mice have been produced by this method, but several laboratories are attempting to develop embryonic stem cells for each of the livestock species. If ES cells can be established for farm animals it may be possible to transfer nuclei from them into enucleated oocytes to produce progeny, thus avoiding the formation of chimera as is required in mice.

Design of Genes for Transfer

The general strategy employed by molecular biologists in designing genes for transfer is to combine parts from two or more existing genes to form a different gene. A gene is composed of two major regions known as the structural region and the regulatory region. The structural region primarily contains the DNA coding sequences that specify the structure of the messenger ribonucleic acid (mRNA). The mRNA passes out of the nucleus into the cytoplasm where it binds to ribosomes to direct the synthesis of a particular protein, such as an enzyme, hormone, or other cellular component.

The regulatory region primarily contains the DNA sequences that regulates when, where and how much mRNA will be produced by the structural part of the gene. Gene expression is the term used to denote the formation of a copy of mRNA from the DNA, which will lead to the synthesis of the specific protein. For example, the growth hormone (GH) gene has a structural region that codes for the GH protein and a regulatory region that causes GH to be synthesized in the pituitary gland. In contrast, the regulatory region of the metallothionein (MT) gene causes MT synthesis in liver, kidney and other tissues. Thus, a MT-GH fusion gene is formed by fusing the MT regulatory region to the GH structural region. When the MT-GH gene was used to produce a transgenic animal, GH was synthesized in liver, kidney and other tissues instead of only in the pituitary gland. Palmiter and coworkers *(14)* used the MT-GH fusion gene to produce the GH transgenic "super mouse". The MT regulatory region was selected because synthesis of MT is normally stimulated by heavy metal ions, so the level of gene expression could be increased by adding zinc to feed or water.

The design of the fusion gene is a crucial factor that determines whether the gene functions correctly in a transgenic animal. Unfortunately, our knowledge of gene regulation is insufficient at present to make decisions based entirely on facts. In addition, expression of fusion genes can vary dramatically between different species, and even from animal to animal within a species. A notable example is the difference in the expression of the oMT-oGH fusion gene between transgenic sheep and mice. When oMT-oGH was transferred into sheep the serum concentrations of oGH in three transgenic lambs varied from 3,800 to 23,000 ng/ml without zinc stimulation. However, in transgenic mice the concentration of oGH varied from 2 to 10 ng/ml without stimulation and 12 to 2,000 ng/ml with zinc stimulation *(15,16)*.

The various genes that have been transferred into farm animals are shown in Table II. Most of the genes that have thus far been transferred were fusion genes, however, a few were the intact gene from a foreign species. The primary

TABLE II. Gene Constructs Integrated into the Genomes of Farm Animals[a]

Gene constructs (regulatory-structural)	Abbreviation[b]	Animal[c]	(Ref.)
Growth Related Genes			
Albumin-Growth Horm. Releasing Fact.(GRF)	mALB-hGRF	P	(17)
Cytomegalovirus(LTR)-Growth Hormone (GH)	CMV-pGH	P	(18)
Mammary Tumor Virus(LTR)-GH	mMTV-bGH	C	(19)
Metallothionein-GH	mMT-hGH	P,R,S	(1,20)
	mMT-bGH	P,S	(21,22)
	oMT-oGH	S	(15)
	hMT-pGH	P	(23)
Metallothionein-GRF	mMT-hGRF	P,S	(22,24,25)
Metallothionein-IGF-I	mMT-hIGF-I	P	(24)
Moloney Leukemia Virus (LTR)-GH	mLV-rGH	P	(26)
	mLV-pGH	P	(18)
Mouse Sarcoma Virus (LTR)-cellular SKI	mSV-cSKI	C,P	(27,28)
Phosphoenolpyruvate Carboxykinase-GH	rPEPCK-bGH	P	(29)
Prolactin-GH	bPRL-bGH	P	(30)
Skeletal Actin-Estrogen Receptor	cASK-hER	C	(31)
Skeletal Actin-IGF-I	cASK-hIGF-I	C	(32)
Transferrin-GH	mTF-bGH	P,S	(7,33)
Disease- & Immunologically-Related Genes			
Immunoglobulin Heavy Chain-*c-myc*	rbEμ-rb*c-myc*	R	(34)
Visna Virus (LTR) & Envelope	VV-ENV	S	(35)
Metallothionein-Mx (influenza resistance)	hMT-mMx	P	(36)
Mouse Mx-Mx	mMx	P	(36)
Sarcoma Virus(LTR)-Mx	mSV-mMx	P	(36)
Immunoglobulin A (α & κ)	mIgA	P,S	(37)
Immunoglobulin G (γ & κ)	mIgG	P,R	(38)
Mammary- or Blood-Specific Gene Expression			
α-S₁casein-Lactoferrin	bαCAS-hLF	C	(39)
β-casein-Tissue Plasminogen Activator	bβCAS-hTPA	G	(40)
ß-lactoglobulin-Factor IX	oßLG-hFIX	S	(41)
ß-lactoglobulin-α-1-anti-trypsin	oßLG-hα1AT	S	(42)
Mammary Tumor Virus(LTR)-IGF-I	MTV-hIGF-I	C	(32)
Whey Acidic Protein	mWAP	P	(43)
Whey Acidic Prot.-Tissue Plasmiogen Activator	mWAP-hTPA	G	(41)
Whey Acidic Protein-Protein C	mWAP-hPC	P	(44)
α-globin-β-globin	hαGLO-hβGLO	P	(45)
Metabolic Pathway Genes			
Metalothioneine-Serine Transacetylase (ST)	oMT-*cysE*	S	(46)
Metalothioneine-O-Acetyl. Sulfhydry. (OAC)	oMT-*cysK*	S	(46)
Rous Sarcoma Virus (LTR)-ST & OAC	RSV-*cysE-cysK*	S	(47)

[a] Adapted from ref. 48. [b] Lower case letters designate species from which DNA sequence was derived: b, bovine; c, chicken; h, human; m, murine; o, ovine; p, porcine; r, rat. [c] C = cattle, G = goat, P = pig, R = rabbit, S = Sheep.

reasons for transferring genes into farm animals are to: improve productivity traits, enhance animal health, produce new biomedical products, and create animal models for human diseases. The current status of research on transgenic farm animals for the first three of these areas will be presented in the following sections. Use of farm animal transgenics as models for human diseases is just beginning.

Improved Productivity Traits

Transfer of genes for improved animal productivity traits, such as feed conversion, rate of gain, reduction of fat, and improved quality of meat, milk, or wool, would have a dramatic impact on the livestock industry and reduce the cost of animal products for the consumer. These productivity traits are controlled by numerous genes, but most of these genes have not yet been identified. The expansion of research on mapping the genomes of livestock species should be helpful in elucidating more of the economically important genes. In addition, the expansion of our knowledge of gene regulation may permit some of the genes that are known to influence productivity to be expressed more appropriately in the future.

Growth Hormone (GH). Numerous GH transgenes have been transferred into livestock species (Table II). Most of the pigs and lambs that expressed these transgenes had continuously elevated GH in their plasma. The concentration of GH varied greatly among transgenics with the same structural gene, which is thought to be the result of random insertion of transgenes into the genome. Plasma concentrations at birth ranged from 3 to 949 hGH ng/ml and 5 to 944 bGH ng/ml in MT-hGH and MT-bGH transgenic pigs, respectively (*1,49*). Pigs expressing the hGH transgene rarely had detectable concentrations of plasma pGH (*49*), which indicates the negative feedback mechanism was functioning. Furthermore, insulin-like growth factor-I (IGF-I) concentrations in hGH and bGH expressing transgenic pigs were three-fold higher than in littermate control pigs.

The effects of bGH gene expression on concentrations of glucose and several metabolic hormones in the plasma are summarized in Table III. In comparison to littermate controls, pigs expressing MT-bGH had significantly elevated levels of glucose and insulin, significantly lower levels of thyroxin and prolactin, and concentrations of cortisol and triiodothyronine were similar. These results are comparable to those reported for pigs injected daily with exogenous pGH, which had average increases in serum glucose ranging from 8% to 48% and concentrations of serum insulin that were 2- to 7-fold higher than in control pigs (*50-52*). Additionally, Ebert and coworkers (*26*) reported that a transgenic pig expressing rat GH had glucosuria and consistently had serum glucose levels more than three-fold higher than control pigs.

Founder MT-hGH and MT-bGH transgenic pigs did not gain faster than their littermate controls, which was probably the consequence of being fed a diet containing only 16% protein. Studies using pigs injected with exogenous pGH indicate that maximal growth rate is attained only if the diet contains adequate protein and, particularly, lysine (*53,54*). Subsequently, when higher levels of dietary protein and lysine were fed during the 30 to 90 kg growth period the G2 and G3 progeny of MT-bGH transgenic founder 37-06 gained weight 11.1% faster,

Table III. Glucose and Metabolic Hormone Concentrations in Plasma
of MT-bGH Transgenic and Littermate Control Pigs

		Control		Transgenic	
Item	N	(mean ± SEM)	N	(mean ± SEM)	P
Glucose[a] (mg/dl)	10	72 ± 5	10	109 ± 13	0.011
Cortisol[b] (ng/ml)	6	39 ± 9	8	37 ± 7	0.84
Insulin[a] (pg/ml)	10	24 ± 4	10	480 ± 118	0.001
Prolactin[a] (ng/ml)	10	3.9 ± 0.5	10	2.3 ± 0.4	0.021
T_3[bc] (ng/ml)	6	1.2 ± 0.3	7	1.3 ± 0.2	0.68
T_4[bd] (ng/ml)	6	49 ± 4	7	29 ± 3	0.003

SOURCE: Reprinted with permission from ref. 7.
[a] Blood collected after overnight fast.
[b] Blood collected from cannulated pigs.
[c] T_3 = triiodothyronine
[d] T_4 = thyroxine

and G2 progeny of 31-04 founder gained 13.7% faster than sibling control pigs
(Table IV).

Elevated concentrations of GH in pigs expressing MT-hGH and MT-bGH
transgenes have produced marked repartitioning of nutrients away from subcutan-
eous fat and into other carcass components, including muscle, skin, bone, and
certain organs. Carcass fat was the most dramatically altered component of MT-
bGH transgenic pigs. As shown in Table V, the difference in the percentage of
carcass fat in transgenic and sibling pigs became more pronounced as they
approached market weight. As the amount of fat declined in MT-bGH transgenic
pigs, the proportion of lean muscle increased (64.8% versus 56.9% separable lean,
P < .01) compared to sibling control pigs (56). The weights and circumferences of
humerus, femur and radius-ulna of bGH transgenic pigs were significantly greater
than for sibling control pigs (24). In contrast, linear length of the humerus and
femur was the same for transgenic and sibling pigs. The adrenal, kidney, liver,
thyroid, and heart were all significantly heavier in bGH transgenic pigs than
siblings, while the weights of remaining organs were similar for both groups (24).

In contrast to the transgenic pigs, transgenic lambs did not grow faster or utilize
feed more efficiently than control lambs, but they were much leaner. In transgenic
lambs, the lack of body fat may have been the result of hyperglycemia and
glycosuria (22,33).

Pigs expressing either the MT-hGH or MT-bGH transgene exhibited several
notable health problems, including lameness, susceptibility to stress, gastric ulcers,
parakeratosis, lethargy, anestrus in gilts, and lack of libido in boars (7,21).

Table IV. Average Daily Gain and Feed Efficiency of MT-bGH Transgenic Pigs (30 to 90 kg Body Weight)

Line	Group[a]	Average daily gain (g ± S.E.M)	Kg feed/kg gain ±S.E.M.
37-06[b]	Control	813 ± 17 (23)[c]	2.99 ± 0.12 (8)
	Transgenic	903 ± 23 (13)	2.46 ± 0.16 (5)
		P = 0.002	P = 0.026
31-04[d]	Control	869 ± 43 (7)	ND [e]
	Transgenic	988 ± 62 (7)	ND
		P = 0.15	

SOURCE: Reprinted with permission from ref. *7.*
[a] Pigs were fed corn-soybean diet containing 18% crude protein plus 0.25% lysine.
[b] G2 and G3 progeny of founder 37-06.
[c] Number of animals indicated in parenthesis.
[d] G2 progeny of founder 31-04.
[e] Not determined because pigs were group fed.

Pathology in joints, characteristic of osteochondritis dissecans, was also reported in a pig expressing a rGH transgene *(26)* and in some pigs treated with exogenous pGH for 57 days *(52).* In contrast, no increase in the incidence of these pathological conditions was observed in non-expressing MT-hGH or MT-bGH transgenic pigs *(21)* or in transgenic pigs that expressed only low levels of bGH *(30)*.

Many of the health problems observed in pigs exposed to high concentrations of GH are quite prevalent in the general swine population but at a lower incidence and with less severity. Several necropsy surveys indicate gastric ulcers were present in 10 to 30% of market hogs at slaughter *(57)*, and up to 90% of rapidly growing pigs have lesions of osteochondrosis, which leads to degenerative joint disease in certain pigs and is the major cause of lameness in swine *(58,59)*. Additional investigation is required to determine whether these ailments would be less prevalent in pigs expressing GH transgenes if the genetic base was rigidly selected for a low incidence of these conditions.

Reproductive capacity was seriously impaired in pigs expressing either the MT-hGH or MT-bGH transgene. Gilts failed to exhibit estrus, and their ovaries were devoid of corpora lutea or corpora albicans when examined at necropsy (V.G. Pursel, unpublished data). Boars totally lacked libido; therefore, spermatozoa were recovered by electroejaculation or were flushed from the epididymis at necropsy to use for artificial insemination to obtain germ line transmission of the transgene *(21)*.

Table V. Percentage of Total Carcass Lipid for MT-bGH Transgenic
and Sibling Control Pigs at Increasing Body Weights

	Body weight (kg)				
Group	14	28	48	68	92
Control	10.0	12.3	16.6	26.8	29.1
MT-bGH	6.2	7.6	8.2	6.0	4.5

SOURCE: adapted from ref. 56.
N = 4 to 6 per weight group.

A major difference between a transgenic pig with elevated GH and a normal pig injected daily with exogenous GH is that in the latter case, GH is elevated episodically, while in the transgenic pig, GH is elevated continuously (49). We hypothesize that continuous exposure to elevated levels of GH contributes to the multiple health problems observed in our MT-hGH and MT-bGH transgenic pigs and also may prevent them from growing to their full potential. There is evidence for the latter in the rat since Robinson and Clark (60) reported that rats infused continuously with a high concentration of GH do not attain the maximal rate of growth achieved in rats injected once daily.

In transgenic mice, both MT and PEPCK regulatory sequences could be manipulated by dietary changes to modify expression of GH transgenes, but these regulatory sequences were much less responsive to dietary manipulation in transgenic pigs and sheep (15,29,61). Use of regulatory regions that permit expression of GH fusion genes only during the rapid growth phase or that can induce the release of large episodic doses of GH may be essential to achieve only the positive aspects of elevated GH for pigs. Several laboratories are continuing to conduct research on this problem because of the obvious economic gain that producers and consumers would obtain from improved efficiency in feed utilization and reduced fat content.

Growth Hormone Releasing Factor (GRF). Transgenic pigs and sheep have been produced using MT and ALB regulatory sequences. Only two of seven (29%) pigs and one of seven (14%) of lambs expressed the MT-hwF transgene (7,22) In contrast, 11 of 14 (79%) of transgenic mice expressed MT-hGRF (62). Reasons for this discrepancy in incidence of expression among species are unknown, but possibly particular combinations of regulatory and structural sequences are less effective in some species than in others. In contrast, all three pigs and two of four lambs expressed the ALB-hGRF transgene. The concentration of GRF in plasma of transgenic pigs with MT-hGRF or ALB-hGRF were 130 to 380 pg/ml and 400 to 8000 pg/ml, respectively (17,24). These values are 10- to 500-fold higher than concentrations of GRF in plasma of littermate control pigs. However, most of the assayable GRF in plasma of the MT-hGRF pigs was the 3-

44 metabolite rather than the native peptide (1-44), which may explain why the concentration of pGH in plasma was not elevated in the transgenics compared to the littermate controls (*17*). In contrast, the transgenic lambs had elevated GH in response to the hGRF transgenes, and as a consequence, the lambs were phenotypically similar to the bGH and oGH transgenic lambs.

Insulin-Like Growth Factor-I (IGF-I). Four transgenic pigs (*24*) and six transgenic calves have been produced with IGF-I transgenes (*32*). However, only one of the pigs expressed elevated levels of IGF-I and it died before growth performance could be evaluated. No information has been published regarding expression or performance of the transgenic calves. Additional research is in progress with IGF-I fusion genes in which expression will be targeted to skeletal muscle.

Stimulation of Muscle Development. Sutrave and coworkers (*63*) reported that mice expressing a chicken cSKI transgene exhibited a distinct phenotype characterized by hypertrophy of skeletal muscles and reduced body fat. The gene transferred into mice consisted of a mouse sarcoma virus (MSV) LTR regulatory sequence fused to a truncated cSKI cDNA. The transgene product is a protein containing 448 amino acids that is localized primarily in muscle nuclei. The normal function of cSKI and its mode of action is unknown.

The MSV-cSKI gene has now been transferred into the genome of swine (*28*) and cattle (*27*). Expression of the cSKI transgene in swine resulted in a wide range of phenotypes among animals. Five transgenic pigs exhibited varying degrees of muscular hypertrophy that was visually detected around 3 months of age. In three pigs, both hams and shoulders appeared enlarged, while in two pigs, hypertrophy was evident only in the shoulders. Levels of gene expression in muscles of these pigs have not yet been determined. In contrast, between birth and 3 months of age, five other cSKI transgenic pigs exhibited muscular atonia and weakness in both the front and rear legs. Skeletal muscles from these pigs had high levels of cSKI mRNA, while cardiac muscle contained low levels, and no transgene mRNA was detected in any other tissue. Histological examination of skeletal muscles from these myopathic pigs revealed that muscle fibers contained large vacuoles. None of the cSKI transgenic mice exhibited the myopathic phenotype found in pigs. However, one hypertrophic line of cSKI mice had centrally located nuclei in some muscle fibers, and other fibers contained small vacuoles.

Muscle phenotype was unaltered in the other cSKI transgenic pigs, seven of which were biopsied and cSKI mRNA evaluated. Northern analysis revealed cSKI expression in biceps femoris of three pigs, in semimembranosus of four pigs, and in triceps brachii of five pigs. More complete characterization of the cSKI transgenic pigs will be made with progeny of these founders when they become available.

The single cSKI transgenic bull calf was phenotypically normal until 8 weeks of age (*27*). Over the next 2 weeks he developed muscle hypertrophy that was most evident in the loin and hind quarters. From 10 to 15 weeks of age the calf's muscles progressively degenerated to the point requiring euthanasia. Histopathological evaluation of muscles revealed enlarged muscle fibers with some fibers undergoing degeneration. High concentrations of cSKI mRNA were present in all six skeletal muscles that were tested.

Wool Production. Cysteine is the rate limiting amino acid for the production of wool. Addition of cysteine to the diet does not increase wool production because the rumen degrades proteins and sulphur is lost as hydrogen sulphide. Bacterial genes are capable of synthesizing cysteine from the hydrogen sulphide; thus, the transgenic approach might be used to introduce the bacterial enzymes into sheep to improve the growth of wool (*64*). Two laboratories in Australia are making rapid progress on producing transgenic sheep with the bacterial genes for serine transacetylase and o-acetylserine sulfhydrylase, which have been isolated and characterized for *Escherichia coli* (*65*) and *Salmonella typhimurium* (*47*). Ward and Nancarrow (*46*) have succeeded in transferring these genes into mice and obtained high levels of active enzyme in the intestine. Rogers and co-workers (*47,66*) have transferred the bacterial genes into both mice and sheep. Only low levels of the bacterial enzymes have been obtained so far in the rumen epithelium of the transgenic sheep. Current research is directed towards evaluating regulatory sequences that may provide more active and specific expression of these bacterial enzymes in the rumen epithelium.

Improved Animal Health

Application of transgenic technology holds considerable promise for improved animal health in the future. Several approaches under investigation include transfer of genes for naturally occurring disease resistance, for preformed antibodies, and for viral envelope proteins. When immunologists have a fuller understanding of the major histocompatibility complex, these genes may be extremely useful for enhancing disease resistance.

Naturally Occurring Disease Resistance. Mice carrying the autosomal dominant Mx1 allele are resistant to influenza virus. This gene has been cloned and characterized (*67*). Interferon stimulates Mx protein production that promotes resistance to viral infection (*68*). Brem (*36*) transferred three Mx fusion genes (Table II) into swine to test their effectiveness. Two of five transgenic pigs harboring the Mx regulatory and structural sequences were found to respond to interferon induction of Mx mRNA. However, the response was insufficient to produce detectable amounts of Mx protein in the tissues. The other two fusion genes that were transferred into pigs were rearranged during integration, so they were not functional.

Preformed Antibodies. Genes encoding mouse α heavy and κ light chains from antibodies against phosphorylcholine (PC) were co-injected into ova to produce two transgenic pigs and three transgenic lambs (*37*). In the transgenic pigs, the mouse immunoglobulin A (IgA) was detected in the serum despite the failure of an intact mouse κ transgene to integrate. Transgenic progeny from both founders demonstrated high levels of serum mouse IgA starting at about 5 weeks of age.

Average levels of mouse IgA were 630 μg/ml in one line and 1293 μg/ml in the other. In both cases, IgA levels in progeny were higher than in the founders. However, mouse IgA showed little binding specificity for PC, presumably because secreted chimeric antibody included endogenous light chains with mouse heavy chains. In transgenic sheep, mouse IgA was detectable in peripheral lymphocytes but not in serum. These studies need to be expanded to obtain conclusive proof that the IgA transgene would be protective against pathogenic bacteria.

In a similar study, Weidle and coworkers (*38*) produced two transgenic pigs and three transgenic rabbits that harbored mouse λ heavy and κ light chain transgenes from antibodies directed against the hapten 4-hydroxy-3-nitrophenylacetate. Titers of 100 to 300 μg IgG/ml in transgenic rabbits and up to 1,000 μg IgG/ml in one transgenic pig were present in the serum of founders and transgenic progeny. Further evaluation of the antibody composition indicated xenogeneic antibodies had formed by association of light chains of rabbit and pig with heavy chains of the mouse.

It is clear from these recent studies that further investigations should consider using homologous regulatory sequences to inhibit formation of chimeric antibodies with low binding specificity for the target antigen.

Enhanced Disease Resistance.   A transgenic approach may be effective for producing farm animals that are genetically resistant to specific pathogenic viruses. Salter and Crittenden (*69*) produced transgenic chickens that were highly resistant to infection with subgroup A avian leukosis virus (ALV) by introducing an ALV gene that encoded a viral envelope glycoprotein. Normally, ALV enters chicken cells by attachment of the envelope glycoprotein to cell membrane receptors. However, in the transgenic chickens the subgroup A ALV virus could not enter the cells because the membrane receptors were presumably occupied with envelope protein that had been produced by the transgene. These chickens were not resistant to infection by subgroup B ALV because a different receptor is used for entry.

A similar experiment is being conducted in sheep with the envelope gene from visna virus. The sheep population throughout the world is widely infected with visna virus, which is an ovine lentivirus similar to equine infectious anemia virus, caprine-arthritis encephalitis virus, and bovine, feline, simian and human immunodeficiency virus. Visna viruses are usually transmitted to lambs in colostrum or milk, where they infect the macrophages or monocytes, and establish life-long infections. The clinical disease in sheep is ovine progressive pneumonia, arthritis, mastitis, and occasionally paralysis (*70*). Immunizations with vaccines have not been effective in control of visna virus or other leniviruses.

Three transgenic sheep have been produced by microinjection of a visna virus fusion gene into pronuclei of sheep zygotes (*35*). The fusion gene was composed of visna virus LTR regulatory region fused to the coding region for visna virus envelope protein. All three lambs expressed the envelope glycoprotein in the macrophages as well as in fibroblasts isolated from the skin of the transgenic lambs. These animals have remained healthy and expression of the viral gene has had no observable detrimental effect. Two of the three sheep are producing antibodies to the envelope protein, which possibly indicates the viral gene was expressed relatively late in development and was not recognized as self-antigen. These sheep have not been challenged with virulent visna virus to test their abilities to resist infection. Such experiments will be conducted in the future with transgenic progeny from these ewes.

New Biomedical Products

Transgenic livestock may soon play an important role in providing new life-saving medical products for treating a variety of human diseases. This is certainly not a new role for farm animals. Thousands of people have in the past benefitted from biomedical products derived from farm animals. Notable examples include replacement heart valves, insulin to treat diabetes, and oxytocin to induce labor during childbirth.

The most significant recent advance in gene transfer involves the direction of expression of considerable quantities of foreign protein to the mammary glands of livestock. A key factor responsible for this high level of transgene expression is the use of genomic DNA for the structural gene instead of cDNA, which had been used previously. These findings fully support the earlier research of Brinster and coworkers (71) in mice that showed inclusion of introns in gene constructs usually resulted in higher levels of transgene expression than when the same regulatory sequences were ligated to cDNA.

The regulatory sequences that have been investigated for their potential to direct expression of foreign proteins to the mammary gland include sheep ß-lactoglobulin (BLG), cow α-S1-casein, rabbit, rat and cow β-casein, and mouse whey acidic protein (WAP). Since the casein proteins comprise the majority of protein content in milk, one might expect casein regulatory sequences to be the most effective for transgenes. So far, this assumption has been incorrect, possibly because several casein genes are clustered together, which may make their regulation more complex than for single genes.

The regulatory sequences that prove to be most effective for directing high levels of a pharmaceutical should also prove to be useful for modifying milk composition for agricultural purposes and vice versa. Several of these non-pharmaceutical purposes include: reduction of mastitis, alteration of casein composition for production of cheese or for production of human proteins in milk of farm animals to provide human infants with a better substitute for a mother's breast milk, alteration of lactose composition, and reduction or alteration in butterfat content (72-74).

Whey Acidic Protein (WAP). The mouse WAP gene was transferred into swine and sheep to test whether this gene, which is normally only found in rodent species, might be effective for expressing transgenes in the mammary glands of livestock (43). Milk has been evaluated from six lines of transgenic pigs and two transgenic sheep. Mouse WAP made up about 3% of the total milk proteins in the transgenic sows and ewes, thus demonstrating that it is possible to produce high levels of a foreign protein in milk of pigs and sheep.

Quite unexpectedly, five sows from three lines of WAP transgenic pigs produced milk only a few days before becoming agalactic (5). Lactation failure was not accompanied by elevated body temperature, lack of maternal behavior, loss of appetite, or by other symptoms that normally accompany agalactia in swine. Neither mice nor sheep expressing other mammary-specific transgenes have previously exhibited this phenomenon. Recent data from mice and pigs indicates that in agalactic females the transgene starts producing WAP early in gestation rather than late in gestation as is common in rodents. This early expression in some way interferes with normal mammary development and causes early involution (*75, 76*).

Human Protein C (hPC). Velander and coworkers (*44*) inserted cDNA of hPC into the first exon of the WAP gene to produce several transgenic pigs. Lactating transgenic sows produced up to 1 g hPC per liter of milk. The biological activity of the recombinant hPC was equivalent to that of protein C derived from human plasma, which makes it commercially feasible to extract hPC from milk, purify it, and clinically evaluate its efficacy and safety.

Protein C plays an important role in the regulation of hemostasis. When sufficient quantities are available, it is anticipated that protein C will be used to treat people who are hereditarily deficient for protein C and, as a consequence, have recurrence of thrombosis (intravascular blood clots). Protein C should also be useful in preventing blood clot formation in septic shock patients and those who have surgical operations such as elective hip replacement. Once a dependable source and clinical effectiveness of hPC have been established, the estimated U.S. market could run as high as 96 kg annually with a market value of $960 million, according to projections of the American Red Cross.

Tissue Plasminogen Activator (hTPA). The mouse WAP regulator has also been used to express a variant of hTPA cDNA in a transgenic goat (*40*). Tissue plasminogen activator (TPA) is an anti-clotting agent currently used to treat patients immediately after suffering a heart attack. Only low concentrations of hTPA were found in the milk, and expression of the hTPA had no adverse effect on milk production or general health of this transgenic goat. However, when another transgenic goat was produced with the β-casein regulatory region instead of the WAP regulatory region, a much higher level of hTPA was produced. This goat became agalactic soon after parturition. Whether the agalactia was the result of expression of hTPA or was just a characteristic of this particular goat remains to be determined by additional experiments (K.M. Ebert, personal communication, 1993).

Human Alpha-1-Antitrypsin (hα_1AT). High levels of hα_1AT expression in milk were obtained in four transgenic ewes in which sheep BLG regulatory sequences were ligated to hα_1AT genomic sequences (*77*). The concentration of hα_1AT at the seventh week of lactation varied from 1.5 to 37.5 g/l among the ewes, and, most remarkably, hα_1AT made up more than half of the protein contained in the milk from one transgenic ewe (*77*). In spite of the extraordinarily high levels of hα_1AT production, the ewes had normal durations of lactation and exhibited no ill effects from the transgene. The high level of transgene expression with genomic hα_1AT sharply contrasted earlier investigations in which only low levels of expression were obtained when BLG had been ligated to cDNA of hα_1AT (74) or human Factor IX (41).

The glycosylation and biological activity of hα_1AT purified from transgenic sheep milk appeared to be indistinguishable from that of hα_1AT purified from human plasma (77). Whether the sugar composition of the carbohydrate side-chains have been altered remains to be determined.

The hα_1AT is being investigated for its potential to treat the large number of people afflicted with a common hereditary deficiency in hα_1AT that predisposes them to life-threatening emphysema. Approximately 20,000 people in the U.S. have this deficiency. At the present time hα_1AT is extracted from human blood plasma, but the large quantity needed per patient (about 200 grams per year) makes that source inadequate and expensive. Currently, a year's supply of hα_1AT costs about $22,000 per patient.

Human Lactoferrin (hLF). A transgenic bull calf was produced with a fusion gene composed of bovine αS_1-casein regulatory region and hLF coding sequences (39). The αS_1-casein portion of the transgene was composed of 15 kb of 5'-flanking and 6 kb of 3'-flanking sequences, which contain elements that have previously conferred tissue-specific expression in transgenic mice (78). It will be several years before the progeny of the transgenic bull can be tested for presence of hLF in milk.

The intended use of hLF is for supplementation of human infant milk formulas. Human breast milk contains more than 2 g lactoferrin per liter. Lactoferrin has bacteriostatic properties that are believed to play an important role in defending infants against gastrointestinal infections. In addition, lactoferrin is a major iron-binding protein in milk, so it may provide the infant with a source of absorbable iron as lactoferrin passes through the digestive system.

Human Hemoglobin (hα-ßGLO). Hemoglobin is one of several biomedical proteins that cannot be synthesized by the mammary gland but could be produced in other organs of transgenic animals and recovered from the blood. In crisis treatment, in the field, free hemoglobin would be superior to whole blood or concentrated red blood cells for transfusions because free hemoglobin does not require refrigeration. Hemoglobin would be particularly useful on the battlefield and for major natural disasters because, unlike red blood cells, it lacks antigenic components. Thus, hemoglobin would be compatible with all blood types.

Scientists at the DNX Corporation, Princeton, New Jersey, have reported producing 11 transgenic pigs that harbor the human alpha and beta globin genes (45,79). In these transgenic pigs, up to 24% of the red blood cells are producing human hemoglobin instead of pig hemoglobin. The hemoglobin is extracted from the red blood cells, and the human and pig hemoglobin are then separated by ion exchange chromatography. The purified hemoglobin must then be chemically modified so that the protein chains are cross linked or polymerized. This chemical alteration is critical; otherwise, the globin chains would be unstable and incapable of releasing oxygen to the tissues when used for a transfusion. Previous clinical trials with chemical modification of human hemoglobin, which had been extracted from human donor blood cells, proved to be unsuccessful. Thus, scientists at

DNX do not expect the human hemoglobin produced from these particular transgenic pigs to be useful for transfusion. Information gained from studying them will be extremely useful for designing subsequent experiments. The next step is to alter human globin genes in such a way that the globin genes are cross linked and have reduced oxygen affinity. A number of research laboratories have research in progress on this complex problem.

Organs for Transplantation. Several commercial organizations have initiated research to genetically engineer swine to provide universally acceptable organs for transplantation into humans. When organs are transplanted between widely differing species the complement system is immediately activated resulting in hyperacute rejection of the xenograft. A possible solution to hyperacute rejection is to transfer the human complement regulatory proteins, such as decay accelerating factor (DAF), which can inhibit complement activation. Transgenic pigs that express a human complement inhibitory protein have been produced by Alexion Pharmaceuticals (USA), Imutran (England), and Nextran Corporation (USA). A preliminary report indicated expression of human DAF was detected on the surface of pig lymphocytes that was able to down regulate human complement activation (*80*). High levels of complement inhibitor expression were found on pig endothelial cells of both capillaries and large blood vessels. Results of organ transplantation from these transgenic pigs into primates have not yet been reported.

Problems Remain. Even though several human therapeutical proteins have now been successfully produced in milk and blood of transgenic animals, some difficult problems must be solved before these products are approved for use. Product safety is a large issue. These products will require the same rigorous scrutiny as the products extracted from animal tissue produced by tissue culture or synthesized by recombinant organisms. Products from transgenic animals must be purified to remove all non-human proteins that might cause allergic reactions. In addition, it is still not known whether these complex human proteins are sufficiently similar in structure and biological activity to the natural proteins produced by the human body so that antibodies are not produced. While scientists are confident that these technical and regulatory obstacles can be overcome, few people are willing to predict how long it will take to work out these problems and complete the clinical testing that will be required to obtain approval of regulatory authorities for marketing to the public.

Conclusions

Although the means to apply recombinant DNA technology to farm animals has become a reality in the past decade, many problems still prevent widespread application of this technology for improvement of animal productivity traits. These problems include a lack of worthy gene candidates, a low rate of efficiency in transferring genes, and inability to precisely regulate the time, level, and duration of transgene expression. The success in directing high levels of transgene expression to the mammary gland should stimulate additional efforts to produce rare human pharmaceuticals for patients, and expansion of research to modify milk composition. At present, the transgenic approach for improvement of farm animals for production purposes remains only a hope for the future.

Literature Cited

1. Hammer, R.E.; Pursel, V.G.; Rexroad, C.E. Jr.; Wall, R.J.; Bolt, D.J.; Ebert, K.M.; Palmiter, R.D.; Brinster, R.L. *Nature*, **1985**, *315*, 680-683.
2. Wall, R.J.; Pursel, V.G.; Hammer, R.E.; Brinster, R.L. *Biol. Reprod.* **1985**, *32*, 645-651.
3. Pursel, V.G.; Wall, R.J. *J. Anim. Sci. (Suppl. 1)*, **1989**, *67*, 375-376. (Abstr.).
4. Wilkie, T.M.; Brinster, R.L.; Palmiter, R.D. *Dev. Biol.* **1986**, *118*:9-18.
5. Shamay, A.; Solinas, S.; Pursel, V.G.; McKnight, R.A.; Alexander, L.; Beattie, C.; Hennighausen, L.; Wall, R.J. *J. Anim. Sci.* **1991**, *69*, 4552-4562.
6. Burdon, T.G.; Wall, R.J. *Mol. Reprod. Dev.* **1992**, *33*, 436-442.
7. Pursel, V. G.; Hammer, R.E.; Bolt, D.J.; Palmiter, R.D.; Brinster, R.L. *J. Reprod. Fertil. (Suppl.)* **1990**, *41*, 77-87.
8. Love, J.; Gribbin, C.; Mather C.; Sang, H. *Bio/Technology*, **1994**, *12*, 60-63.
9. Naito, M.; Sasaki, E.; Ohtaki, M.; Sakurai, M. *Mol. Reprod. Dev.* **1994**, *37*, 167-171.
10. Jaenisch, R. *Proc. Natl. Acad. Sci. USA*, **1976**, *73*, 1260-1264.
11. Salter, D.W.; Smith, E.J; Hughes, S.H.; Wright, S.E.; Crittenden, L.B. *Virology*, **1987**, *157*, 236-240.
12. Jaenisch, R. *Science* **1988**, *240*, 1468-1474.
13. Capecchi, M.R. *Science* **1989**, *244*, 1288-1292.
14. Palmiter, R.D.; Brinster, R.L.; Hammer, R.E.; Trumbauer, M.E.; Rosenfeld, M.G.; Birnberg, N.C.; Evans, R.M. *Nature*, **1982,** *300*, 611-615.
15. Murray, J.D.; Nancarrow, C.D.; Marshall, J.T.; Hazelton I.G.; Ward, K.A. *Reprod. Fert. Dev.* **1989**, *1,* 147-155.
16. Nancarrow, C.D.; Shanahan, C.M.; Byrne, C.; Marshall, J.T.A.; Murray, J.D.; Hazelton, I.G.; Rigby, N.W.; Townrow, C.T.; Ward, K.A. *11th Int. Congr. Anim. Reprod. Artif. Insem. Dublin.* **1988**, *5*, 478.
17. Pursel, V.G.; Miller, K.F.; Bolt, D.J.; Pinkert, C.A.; Hammer, R.E.; Palmiter, R.D.; Brinster, R.L. In *Biotechnology of Growth Regulation*; Heap, R.B., Prosser, C.G., Lamming, G.E., Eds.; Butterworths, London, **1989**, pp 181-188.
18. Ebert, K.M.; Smith, T.E.; Buonoma, F.C.; Overstrom E.W.; Low, M.J. *Anim. Biotech.* **1990**, *1*, 145-159.
19. Roshlau, K.; Rommel, P.; Andreewa, L.; Zackel, M.; Roschlau, D.; Zackel, B.; Schwerin, M.; Huhn R.; Gazaejan, K.G. *J. Reprod. Fertil. (Suppl.)* **1989**, *38*, 153-160.

20. Brem, G.; Brenig, B.; Goodman, H.M.; Selden, R.C.; Graf, F.; Kruff, B.; Springman, K.; Hondele, J.; Meyer, J.; Winnaker, E-L.; Krausslich, H. *Zuchthyg.* **1985**, *20*, 251-252.

21. Pursel, V.G.; Rexroad, C.E. Jr.; Bolt, D.J.; Miller, K.F.; Wall, R.J.; Hammer, R.E.; Pinkert, C.A.; Palmiter, R.D.; Brinster, R.L. *Vet. Immunol. Immunopathol.* **1987**, *17*, 303-312.

22. Rexroad, C.E. Jr.; Hammer, R.E.; Bolt, D.J.; Mayo, K.M.; Frohman, L.A.; Palmiter, R.D.; Brinster, R.L. *Mol. Reprod. Dev.* **1989**, *1*, 164-169.

23. Vize, P.D.; Michalska, A.E.; Ashman, R.; Lloyd, B.; Stone, B.A.; Quinn, P.; Wells, J.R.E.; Seamark, R.F. *J. Cell. Sci.* **1988**, *90*, 295-300.

24. Pursel, V.G.; Pinkert, C.A.; Miller, K.F.; Bolt, D.J.; Campbell, R.G.; Palmiter, R.D.; Brinster, R.L.; Hammer, R.E. *Science* **1989**, *244*, 1281-1288.

25. Brem, G.; Brenig, B.; Muller, M.; Kraublich, H.; Winnacker, E.L. *Occasional Publ. Brit. Soc. Anim. Prod.* **1988**, *12*, 15-31.

26. Ebert, K.M.; Low, M.J.; Overstrom, E.W.; Buonomo, F.C.; Baile, C.A.; Roberts, T.M.; Lee, A.; Mandel, G.; Goodman, R.H. *Mol. Endocrinol.* **1988**, *2*, 277-283.

27. Bowen, R.A.; Reed, M.L.; Schnieke, A.; Seidel, G.E. Jr.; Stacey, A.; Thomas, W.K.; Kajikawa O. *Biol. Reprod.* **1994**, *50*, 664-668.

28. Pursel, V.G.; Sutrave, P.; Wall, R.J.; Kelly, A.M.; Hughes, S.H. *Theriogenology*, **1992**, *37*, 278. (Abstr.)

29. Wieghart, M.; Hoover, J.L.; McCrane, M.M.; Hanson, R.W.; Rottman, F.M.; Holtzman, S.H.; Wagner, T.E.; Pinkert, C.A. *J. Reprod. Fertil. (Suppl.)* **1990**, *41*, 89-96.

30. Polge, E.J.C.; Barton, S.C.; Surani, M.H.A.; Miller, J.R.; Wagner, T.; Elsome, K.; Davis, A.J.; Goode, J.A.; Foxroft, G.R.; Heap, R.B. In *Biotechnology of Growth Regulation.*; Heap, R.B., Prosser, C.G., Lamming, G.E., Eds.; Butterworths, London, **1989**; pp 189-199.

31. Massey, J.M. *J. Reprod. Fertil. (Suppl.)* **1990**, *41*,199-208.

32. Hill, K.G.; Curry, J.; DeMayo, F.J.; Jones-Diller, K.; Slapak J.R.; Bondioli, K.R. *Theriogenology,* **1992**. *37*, 222. (Abstr.).

33. Rexroad, C.E., Jr.; Mayo, K.M.; Bolt, D.J.; Elsasser, T.H.; Miller, K.F.; Behringer, R.R.; Palmiter R.D.; Brinster, R.L. *J. Anim. Sci.* **1991**, *69*, 2995-3004.

34. Knight, K.L.; Spieker-Polet, H.K.; Dori, S.; Oi, V.T. *Proc. Natl. Acad. Sci. USA* **1988**, *85*, 3130-3134.

35. Clements, J.E.; Wall, R.J.; Narayan, O.; Hauer, D.; Schoborg, R.; Sheffer, D.; Powell, A.; Carruth, L.M.; Zink, M.C.; Rexroad, C.E., Jr. *Virology* **1994**, *200*, 370-380.

36. Brem, G. *Mol. Reprod. Dev.* **1993**, *36*, 242-244.

37. Lo, D.; Pursel, V.; Linton, P.J.; Sandgren, E.; Behringer, R.; Rexroad, C.; Palmiter, R.D.; Brinster, R.L. *European J. Immunol.* **1991**, *21*, 1001-1006.

38. Weidle, U.H.; Lenz, H.; Brem, G. *Gene* **1991**, *98*, 185-191.

39.   Krimpenfort, P.; Rademakers, A.; Eyestone, W.; Van de Schans, A.; Van den Broek, S.; Kooiman, P.; Kootwijk, E.; Platenburg, G.; Pieper, F.; Strijker R.; de Boer, H. *Bio/technology* **1991**, *9*, 844-847.

40.   Ebert, K.M.; Selgrath, J.P.; DiTullio, P.; Denman, J.; Smith, T.E.; Memon, M.A.; Schindler, J.E.; Monastersky, G.M.; Vitale, J.A.; Gordon, K. *Bio/technology* **1991**, *9*, 835-838.

41.   Clark, A.J.; Bessos, H.; Bishop, J.O.; Brown, P.; Harris, S.; Lathe, R.; McClenaghan, M.; Prowse, C.; Simons, J.P.; Whitelaw, C.B.A.; Wilmut, I. *Bio/technology* **1989**, *7*, 487-492.

42.   Simons, J.P.; Wilmut, I.; Clark, A.J.; Archibald, A.L.; Bishop, J.O. Lather, R. *Bio/technology* **1988**, *6*, 179-183.

43.   Wall, R.J.; Pursel, V.G.; Shamay, A.; McKnight, R.; Pittius, C.; Hennighausen, L. *Proc. Natl. Acad. Sci. USA* **1991**, *88*, 1696-1700.

44.   Velander, W.H.; Johnson, J.L.; Page, R.L.; Russell, C.G.; Subramanian, A.; Wilkins, T.D.; Gwazdauskas, F.C.; Pittius, C.; Drohan, W.N. *Proc. Natl. Acad. Sci. USA* **1992**, *89*, 12003-12008.

45.   Swanson, M.E.; Martin, M.J.; O'Connell, J.K.; Hoover, K.; Lago, W.; Huntress, V.; Parsons, C.T.; Pinkert, C.A.; Pilder S.; Logan, J.S. *Bio/techology* **1992**, *10*, 557-559.

46.   Ward, K.A.; Nancarrow, C.D. *Experientia* **1991**, *47*, 913-922.

47.   Rogers, G.E. *Trends Biotech.* **1990**, *8*, 6-11.

48.   Pursel, V.G.; C.E. Rexroad, Jr. *J. Anim. Sci. (Suppl. 3)* **1993**, *71*, 10-19.

49.   Miller, K.F.; Bolt, D.J.; Pursel, V.G.; Hammer, R.E.; Pinkert, C.A.; Palmiter, R.D.; Brinster, R.L. *J. Endocrinol.* **1989**, *120*, 481-488.

50.   Campbell, R.G.; Steele, N.C.; Caperna, T.J.; McMurtry, J.P.; Solomon, M.B.; Mitchell, A.D. *J. Anim. Sci.* **1988**, *66*, 1643-1655.

51.   Campbell, R.G.; Steele, N.C.; Caperna, T.J.; McMurtry, J.P.; Solomon, M.B.; Mitchell, A.D. *J. Anim. Sci.* **1989**, *67*, 177-186.

52.   Evock, C.M.; Etherton, T.D.; Chung, C.S.; Ivy, R.E. *J. Anim. Sci.* **1988**, *66*, 1928-1941.

53.   Goodband, R.D.; Nelssen, J.L.; Hines, R.H.; Kropf, D.H.; Thaler, R.C.; Schricker, B.R.; Fitzner, G.E. *J. Anim. Sci.(Suppl. 1)* **1988**, *66*, 95. (Abstr.).

54.   Newcomb, M.D.; Grebner, G.L.; Bechtel, P.J.; McKeith, F.K.; Novakofski, J.; McLaren, D.G.; Easter, R.A. *J. Anim. Sci.(Suppl. 1)* **1988**, *66*, 281. (Abstr.).

55.   McCusker, R.H.; Campion, D.R. *J. Anim. Sci.* **1986**, *63*, 1126-1133.

56.   Pursel, V.G.; Solomon, M.B. *Food Rev. Internat.* **1993**, *9*, 423-439.

57.   O'Brien, J.J. In *Diseases of Swine*; Leman, A.D., Straw, B., Glock, R.D., Mengeling, W.L., Penny, R.H.C. & Scholl, E., Eds.; 6th edition; Iowa State Press, Ames, Iowa, **1986**; pp 725-737.

58.   Reiland, S.; Ordell, N.; Lundeheim, N.; Olsson, S. *Acta Radiol. Suppl.* **1978**. *358*, 123-137.

59.   Carlson, C.S.; Hilley, H.D. Meuten, D.J.; Hagan, J.M.; Moser, R.L. *Am. J. Vet. Res.* **1988**, *49*, 396-402.

60. Robinson, I.C.A.F.; Clark, R.G. In *Biotechnology of Growth Regulation*; Heap, R.B.,Prosser, C.G.,Lamming, G.E., Eds.; Butterworths, London, **1989**; pp 129-140.
61. Pursel, V.G.; Bolt, D.J.; Miller, K.F.; Pinkert, C.A.; Hammer, R.E.; Palmiter, R.D.; Brinster, R.L. *J. Reprod. Fertil. (Suppl.)* **1990**, *40*, 235-245.
62. Hammer, R.E.; Brinster, R.L.; Rosenfeld, M.G.; Evans, R.M.; Mayo, K.E. *Nature* **1985**, *315*, 413-416.
63. Sutrave, P.; Kelly, A.M.; Hughes, S.H. *Genes & Develop.* **1990**, *4*, 1462-1472.
64. Ward, K.A.; Murray, J.D.; Nancarrow, C.D.; Sutton, R.; Boland, M.P. In *Reproduction in Sheep*; Lindsay, D. T.; Pearce, D. T., Eds.; Australian Academy of Science and Australian Wool Corporation, Canberra, Australia, **1984**; pp 279-285.
65. Ward, K.A.; Nancarrow, C.D.; Byrne, C.R.; Shanahan, C.M.; Murray, J.D.; Leish, Z.; Townbow, C. N.; Rigby, M.; Wilson, B.W.; Hunt, C.L. *Rev. Sci. Tec. Off. Int. Epiz.* **1990**, *9*, 847-864.
66. Rogers, G.E.; Sivaprasad, A.V.; Powell, B.C.; Bawden, C.S.; Walker, S.K. *17th Int. Congr. Genet., Birmingham, England* **1993**, pp 89. (Abstr.).
67. Staeheli, P.; Haller, O.; Boll, W.; Lindenmann J.; Weissmann, C. *Cell* **1986**, *44*, 147-158.
68. Muller, M.; Brem, G. *Experientia* **1991**, *47*, 923-934.
69. Salter, D.W.; Crittenden, L.B. *Theor. Appl. Genet.* **1989**, *77*, 457-461.
70. Narayan O.; Cork, C. *Rev. Infect. Dis.* **1975**, *7*, 89-98.
71. Brinster, R.L.; Allen, J.M.; Behringer, R.R.; Gelinas, R.E.; Palmiter, R.D. *Proc. Natl. Acad. Sci. USA*, **1988**, *85*, 836-840.
72. Jimenez-Flores, R.; Richardson, T. *J. Dairy Sci.* **1988**, *71*, 2640-2654.
73. Bremel, R.D.; Yom H.-C.; Bleck, G.T. *J. Dairy Sci.* **1989**, *72*, 2826-2833.
74. Wilmut, I.; Archibald, A.L.; Harris, S.; McClenaghan, M.; Simons, J.P.; Whitelaw, C.B.A.; Clark, A.J. *J. Reprod. Fertil. (Suppl.)* **1990**, *41*, 135-146.
75. Shamay, A.; Pursel, V.G.; Wilkinson, E.; Wall, R.J.; Hennighausen; L. *Transgenic Res.* **1992**, *1*, 124-132.
76. McKnight, R.A.; Shamay, A.; Sankaran, L.; Wall, R.J.; Hennighausen, L. *Proc. Natl. Acad. Sci. USA* **1992**, *89*, 6943-6947.
77. Wright, G.; Carver, A.; Cottom, D.; Reeves, D.; Scott, A.; Simons, P.; Wilmut, I.; Garner I.; Colman, A. *Bio/Technology* **1991**, *9*, 830-834.
78. Meade, H.; Gates, L.; Lacy, E.; Lonberg, N. *Bio/Technology* **1990**, *8*, 443-446.
79. Sharma, A.; Martin, M.J.; Okabe, J.F.; Truglio, R.A.; Dhanjal, N.K.; Logan, J.S; Kumar, R. *Bio/Technology* **1994**, *12*, 55-59.
80. Cozzi, E.; Langford, G.A.; Richards, A.; Elsome, K.; Lancaster, R.; Chen, P.; Yannoutsos, N.; White, D.J.G. *Transplant. Proc.* **1994**. *26*, 1402-1403.

RECEIVED June 30, 1995

INDEXES

Author Index

Abe, K., 124
Alink, G. M., 134
Arai, S., 124
Bartnicki, D. E., 148
Bennett, Alan B., 88
Berg, J. H. J. van den, 134
Bienenmann-Ploum, M. E., 134
Cavadini, C., 181
Chetelat, Roger, 88
Emlay, Donald, 72
Engel, Karl-Heinz, 1,23
Feldman, J., 148
Frick, Oscar L., 100
Fuchs, R. L., 148
Hall, R. L., 59
Hallagan, J. B., 59
Hammes, W. P., 181
Hammond, B. G., 148
Hertel, C., 181
Hiatt, William, 72
Jørgensen, Steen T., 196
Jensen, Bent F., 196
Jones, Daniel D., 52
Kawasaki, M., 171
Keck, P. J., 148
Klann, Ellen, 88
Klein, G., 23
Kuiper, H. A., 134
Kuroda, M., 124

Lavrik, P. B., 148
Love, S. L., 148
Martineau, Belinda, 72
Maryanski, J. H., 12
Matsumoto, I., 124
Murry, Lynn E., 113
Naylor, M. W., 148
Noteborn, H. P. J. M., 134
Pedersen, Sven, 196
Pensa, M., 134
Pursel, V. G., 210
Redenbaugh, Keith, 72
Reynaerts, A., 134
Rogan, G. J., 148
Schauzu, M., 23
Sims, S. R., 148
Somogyi, A., 23
Sone, H., 171
Stahl, U., 160
Takahara, Ryoji, 33
Takahashi, R., 171
Takeoka, Gary R., 1
Teranishi, Roy, 1
Vogel, J., 160
Wackerbauer, K., 160
Watanabe, H., 124
Yamano, S., 171
Young, Alvin L., 52
Zolla, L., 134

Affiliation Index

Agricultural Research Service, 1,210
Agricultural University Wageningen, 134
Bundesinstitut für gesundheitlichen
  Verbraucherschutz und
  Veterinärmedizin, 1,23
Calgene, Inc., 72
Flavor and Extract Manufacturers
  Association, 59

Forschungsinstitut für Brauerei und
  Mälzerei der Versuchs- und Lehranstalt
  für Brauerei in Berlin, 160
Hohenheim University, 181
Hybritech Seed International, Inc., 148
Incyte Pharmaceuticals, 113
Kirin Brewery Co., Ltd., 171
Law Offices of Daniel R. Thompson, 59

Ministry of Health and Welfare (Japan), 33
Monsanto Company, 148
Novo Nordisk A/S, 196
Plant Genetic Systems, 134
RIKILT-DLO, 134
SME Ricerche SCPA, 134
Technische Universität Berlin, 160

U.S. Department of Agriculture, 1,52,210
U.S. Food and Drug Administration, 12
University La Tuscia, 134
University of California—Davis, 88
University of California—San Francisco,100
University of Idaho, 148
University of Tokyo, 124

Subject Index

A

Acetobacter pasteurianus, cloning and analysis of acetolactate decarboxylase gene, 163,165f,167
Acetolactate decarboxylase, function, 172
Acetolactate decarboxylase gene from Acetobacter pasteurianus
cloning and analysis, 163,164f
expression in laboratory yeast strains, 163,165f,167
Acetolactate decarboxylase producing brewer's yeast
analysis of brewed beer
natural toxicants, 176f,177
saccharide and organic acid composition, 177,179
construction
genetic modification procedure, 173,174f
laboratory-scale fermentation test, 173,176f
site, 172–173
formation and removal of diacetyl, 172,174f
future work, 179
objective, 172
safety
characteristics of transformant, 175,178f
DNA analysis of transformant, 175
genetic stability, 177
host and donor organisms, 173
integration of acetolactate decarboxylase expression cassette, 175,176f

Acetolactate decarboxylase producing brewer's yeast—Continued
safety—Continued
introduced gene and resulting gene product, 175
removal of undesirable DNA sequences, 175
Advisory Committee on Novel Foods and Processes, objective, 30–31
Advisory committee role in advising government on agricultural science policy
advantages, 53
Agricultural Biotechnology Research Advisory Committee, 53–57
Agricultural biotechnology, development, 52
Agricultural Biotechnology Research Advisory Committee
accomplishments, 54–55
biosafety results of field tests, 56
composition, 54
establishment, 53
food safety issues, 57
future issues, 57
organic foods produced by biotechnology, 56
purpose, 53–54
role in public acceptance of scientific research, 55–56
societal–bioethical aspects of biotechnology, 57
Agricultural development, use of genetic modification of plants and animals, 88

Agricultural science policy, role of public
and federal advisory committees, 52–57
Agrobacterium rhizogenes, use in food
quality improvement, 4
Agrobacterium tumefaciens, genetic
engineering of plants, 2
Agronomic trait, loss, 121–122
Allergen(s)
characterization, 105
examples, 105–108
functions, 105
ingested food allergens, 105
inhaled allergens, 105
Allergenic potential, potatoes, 155
Allergenicity in transgenic foods
allergens, 105–108
allergy, 102–105
examples, 108–110
transgenic foods, 108–110
Allergy
definition, 102
genetic predisposition, 102–103
late-phase reaction, 104
mechanism, 102–103
occurrence, 104–105
skin test, 103–104
transference to another species through
gene transfer, 105
Aminoglycoside 3'-phosphotransferase II,
safety, 82,84–85
Anaphylaxis
causes, 101
definition, 101
Animal(s)
farm, progress on genetic modifications,
210–225
recombinant DNA techniques, 6
Animal health improvements, genetic
modifications of farm animals, 220–222
Antibiotic resistance marker genes used in
foods produced by recombinant DNA
techniques, safety assessment, 46–48
Antipest effect, cereal cystatins, 129–131
Antisense technique, description, 2
α_1-Antitrypsin, human, genetic
modifications of farm animals, 223–224
Antiviral effect, cereal cystatins, 128t,129

B

Bacillus thuringiensis, use for insect
resistance, 3
Bacillus thuringiensis cotton seed,
transgenic food, 109
Bacillus thuringiensis insecticidal
crystal protein CRYIA(b) expressed in
transgenic tomatoes, safety assessment,
134–145
Bacillus thuringiensis subsp. Itenebrionis
protein, safety assessment in potatoes,
152–153
Barley yellow dwarf virus, role in crop
loss, 115–117
"Basic Principle Document for Food
Biotechnology, Manufacturing and
Safety Assessment Guidelines for Foods
Derived from Recombinant DNA
Techniques", objective, 33
Beer brewing improvement using
genetically modified yeast
acetolactate decarboxylase gene from
Acetobacter pasteurianus
cloning and analysis, 163,164f
expression
in brewer's yeast, 167
in laboratory yeast strains,
163,165f,167
secretion into medium, 167,169
diacetyl content reduction, 161,163
diacetyl production, 161,162f
fermentation
with brewer's yeast, 168f–170
with laboratory yeast strains, 166f,167
safety assessment, 170
Beverages, use of microorganisms in
production, 1–2
Bioengineered foods, safety concerns,
101–102
Biomedical products, genetic modifications
of farm animals, 222–225
Biotechnological food safety in Japan
application of concept of novelty to
foods and food additives produced by
biotechnology, 35–39
application review for confirmation, 34

Biotechnological food safety in Japan—
  Continued
  food allergies as related to foods and
    additives produced by biotechnology,
    48–51
  guideline(s), 33–34
  guideline establishment, 34
  interim report in 1993, 34–35
  safety assessment
    antibiotic resistance marker genes
      used in foods produced by
      recombinant DNA techniques, 46–48
    food crops produced by recombinant
      DNA techniques, 39–46
Biotechnology
  flavor production, 60–61
  food production
    advantages, 51
    application, 51
    importance, 101
Biotechnology-developed foods, Food and
  Drug Administration's policy, 12–21
Bovine somatotropic hormone, usage, 101
Brewer's yeast
  acetolactate decarboxylase producing,
    See Acetolactate decarboxylase
    producing brewer's yeast
  improvements using recombinant DNA
    technology, 171–172
  selection, 171

C

Carbohydrate composition of tomato,
  effect of genetic modification, 89
α-Casein, role as allergen, 106
Castor bean, allergens, 107
Cereal, allergens, 107
Cereal cystatins
  antipest effect, 129–131
  antiviral effect, 128*t*,129
  construction of transgenic rice cultivar,
    131
  molecular cloning, 125–127,129
  safety, 132

Chymosin
  FDA approval, 12
  FDA policy, 13–14
  recombinant, FDA regulation, 61
  recombinant DNA techniques, 5–6
Cloning and analysis, acetolactate
  decarboxylase gene from *Acetobacter
  pasteurianus*, 163,164*f*
Coat protein, protection against, viruses,
  115
Colchicine, use for polyploidization, 114
Colorado potato beetle
  control techniques, 148
  crop damage, 148
  safety assessment of resistant potatoes,
    148–155
Corn viruses, symptoms, 117,118*f*
Cotransformation, use for selectable
  marker problem, 121
Cow's milk, allergens, 105–106
Crustacea, allergens, 108
CRYIA(b) protein
  digestibility upon gastrointestinal
    tract passage in rats, 138
  hemolytic effects, 139
  in vitro binding to gastrointestinal
    tract tissues of mammals, 137
  in vitro degradation, 138
  in vivo binding to gastrointestinal
    tract tissues of rat, 137
  insecticidal activity, 136–137
  short-term toxicity
    in mice, 138–139
    in rabbits, 139
  toxicity profile, 138
Cyclodextrin(s)
  commercial applications, 197
  production, 197
  structures, 197,198*f*
Cyclodextrin glycosyltransferase,
  Thermoanaerobacter, See
  Thermoanaerobacter cyclodextrin
  glycosyltransferase
Cystatins
  cereal, *See* Cereal cystatins
  description, 124

D

Diacetyl
content reduction in beer, 161,163
production, 161,162*f*
Disease-resistant crop production, plant
virus genes, 113–122
DNA techniques, recombinant, *See*
Recombinant DNA techniques

E

Egg, role as allergen, 106
Engineered coat protein protection in
transgenic tobacco, concept, 115
Engineered genes
carbohydrate composition of tomato, 89
genetic engineering of sucrose
accumulation in tomato, 95,96*f*
introgression of sucrose accumulation
from wild tomato species, 90–94
molecular cloning of gene responsible
for sucrose accumulation, 93–95
plant breeding, 90
safety of genetic modifications associated
with sucrose accumulation, 95–97
sources of genetic variation, 90
Environment, role in severity of viral
diseases, 115–117
Enzymes
from genetically modified microorganisms
commercial examples, 196
cyclodextrins, 197,198*f*
safety
cloned organism, 203
donor organism, 203
evaluation principles, 199,202–203
host organism, 203
manufacture and quality control,
199,292*t*
studies, 203–204
safety evaluations, 196–197
Thermoanaerobacter cyclodextrin
glycosyltransferase
development, 197,199–201*f*
protein engineering, 204–207
recombinant DNA techniques, 5–6

European regulation on novel foods and
food ingredients
authorization procedure, 28
criteria for authorization, 27
initial assessment, 27–28
labeling provisions, 28–29
proposal, 26
scope, 27
European Union regulatory framework for
biotechnology
assessment strategies, 29–31
horizontal legislation, 24–26
need, 23
novel foods and ingredients, 26–29
philosophies, 24
vertical legislation, 26
Exotic germ plasm, *See* Engineered genes

F

Farm animals, progress on genetic
modifications, 210–225
Federal advisory committees, role in
providing advice to government on
agricultural science policy, 52–57
Fermentation of food, role of lactic acid
bacteria, 183–185
Field test, biosafety results, 56
Fish, allergens, 107
Flavor
creation for specific needs, 60
definition, 59–60
demands, 59
determinant for food choice, 59
production using modern biotechnology,
60–61
sources, 60
Flavor and Extract Manufacturers
Association GRAS assessment program
for flavor ingredients
analysis by chemical structure
analogy, 62
criteria, 62
establishment, 62
levels of analysis, 63–67
modifications, 62–63
validity, 61–62

Flavor ingredients, Flavor and Extract Manufacturers Association GRAS assessment program, 61–67
Flavor ingredients produced by genetically modified organisms, safety assessment, 59–67
Flavr Savr tomato
description, 2
FDA approval, 12
FDA policy, 17–20
safety assessment, 72–85
transgenic food, 108–109
Food
production via recombinant DNA techniques, 1–8
use of microorganisms in production, 1–2
Food allergies as related to foods and additives produced by biotechnology
documents for applications, 50
future strategies and suggestions, 50–51
labeling, 50
novelty, 50
principles, 48–49
relationship between proteins in recombinants and allergenicity, 49–50
Food(s) and additives produced by biotechnology
concept of novelty, 35–39
food allergies, 48–51
Food and Drug Administration (FDA) policy for biotechnologically developed foods
allergen testing, 16–17
chymosin, 13–14
examples, 21
Flavr Savr tomato, 17–20
foods derived from new plant varieties, 14–17
goal, 17
informal notification process for new plant varieties, 21
labeling, 20
multidisciplinary approach for safety and nutritional composition evaluation, 16
policy statement, 14–15
premarket approval, 16
reference standards, 15–16
role in ensuring food safety, 13

Food and Drug Administration (FDA) regulation, recombinant chymosin, 61
Food biotechnology, role of microorganisms, 160
Food crops produced by recombinant DNA techniques, safety assessment, 39–46
Food(s) cross-reacting with inhalant allergens, role as allergen, 108
Food(s) derived from new plant varieties, FDA policy, 14–17
Food(s) developed by biotechnology
advantages, 12
FDA policy, 13–21
Food fermentation, role of lactic acid bacteria, 183–185
Food ingredients
production via recombinant DNA techniques, 1–8
recombinant DNA techniques, 6
Food or additive identical with host, safety assessment method, 42
Food(s) produced by recombinant DNA techniques, safety assessment of antibiotic resistance marker genes, 46–48
Food production, advantages of biotechnology, 51
Food quality, use of recombinant DNA techniques, 3–4
Fruit quality, improvement of engineered genes, 88–97
Fruit total soluble solids, levels in tomato, 89
Fungi, odors, 160

G

Gene farming, description, 6
Gene splicing, description, 12–13
Generally recognized as safe (GRAS)
assessment program for flavor ingredients, See Flavor and Extract Manufacturers Association GRAS assessment program for flavor ingredients
description, 15
Genetic composition of domestic animals, manipulation, 210
Genetic engineering
description, 12–13

Genetic engineering—*Continued*
development of regulation, 23
Genetic modification
brewer's yeast to produce acetolactate
decarboxylase, 171–179
farm animals
animal health improvements
enhanced disease resistance, 221–222
naturally occurring disease resistance,
220
preformed antibodies, 220–221
biomedical products
development, 222
human α_1-antitrypsin, 223–224
human hemoglobin, 224–225
human lactoferrin, 224
human protein C, 223
organs for transplantation, 225
problems, 225
tissue plasminogen activator, 223
whey acidic protein, 222–223
design of genes for transfer, 213–215
methods of transferring genes
microinjection of pronuclei, 211–212
retroviral insertion, 212
stem cell insertion, 212–213
productivity trait improvements
growth hormone, 215–218
growth hormone releasing factor,
218–219
insulin-like growth factor I, 219
stimulation of muscle development,
219–220
wool production, 220
progress, 210–225
plants and animals, basis of agricultural
development, 88
plants, animals, and microprobes
concerns, 52–53
importance, 52
Genetic strategies, improvement of fruit
quality, 88–97
Genetically engineered crops, safety
assessment, 72–85
Genetically modified food safety
assessment
international organizations, 29
scientific advisory committees, 30–31

Genetically modified foods
need for regulation, 23
regulatory oversight and safety
assessment in European Union, 23–31
Genetically modified lactic acid bacteria,
safety assessment, 181–193
Genetically modified microorganisms,
enzymes, 196–207
Genetically modified organisms
production using modern biotechnology,
60–61
safety assessment of flavor ingredients
produced, 59–67
Genetically modified yeast, beer brewing
improvement, 160–170
Green revolution, description, 88–89
Growth hormone, genetic modification of
farm animals, 215–218
Growth hormone releasing factor, genetic
modification of farm animals, 218–219
Guidance to industry, description, 15

H

Hemoglobin, human, genetic modifications
of farm animals, 224–225
Herbicide tolerance, use of recombinant
DNA techniques, 2–3
Horizontal legislation
authorization procedure, 25–26
directives, 24–25
goal, 24
Human life, relation to lactic acid
bacteria, 182–183

I

Ingested food allergens, characterization,
105
Inhaled allergens, characterization, 105
Insect resistance, use of recombinant DNA
techniques, 3
International organizations, genetically
modified food safety assessment, 29
Introduced gene, safety assessment method,
42–44

J

Japan, biotechnological food safety, 33–51
Joint FAO/WHO Consultation, genetically
modified food safety assessment, 29

K

Kanamycin resistance marker gene,
safety, 85

L

Labeling, FDA policy, 20
Lactic acid bacteria
genetically modified, safety assessment,
181–193
relation to human life, 182–183
role in food fermentation, 183–185
use of genetic modification, 181–182
Lactoferrin, human, genetic modifications
of farm animals, 224
β-Lactoglobulin, role as allergen, 106
Lactones, flavor and odor, 160
Legume, allergens, 106–107
Lemon geranium, use in food quality
improvement, 4
Leptinotarsa decemlineata, See Colorado
potato beetle

M

Maillard adducts, role as allergen, 106
Maize dwarf mosaic virus, role in crop
loss, 117,119,120*f*
Marker genes used in foods produced by
recombinant DNA techniques, antibiotic
resistance, safety assessment, 46–48
Member States of the European Union,
genetically modified food safety
assessment, 29
Microinjection of pronuclei, method of
transferring genes, 211–212
Microorganisms
genetically modified, safety of enzymes,
196–207
recombinant DNA techniques, 4–5

Microorganisms—*Continued*
role in food biotechnology, 160
use in production of foods and
beverages, 1–2
Modern biotechnology, scope, 1
Modified soybean, transgenic food,
109–110
Molecular cloning, cereal cystatins,
125–127,129

N

Netherlands, scientific advisory
committees for genetically modified
food safety assessment, 30
New plant variety foods, Food and Drug
Administration's policy, 14–17
No material difference, description, 149
Novelty of foods and food additives
produced by biotechnology
applicability, 35–36
demonstration of substantial
equivalence, 36
modification of animals and
microorganisms by DNA techniques, 39
principle of foods or additives
identical with conventional ones, 37–39
principle of foods or additives
identical with existing ones, 37
NPTII protein
in vitro degradation, 138
safety assessment in potatoes, 153
Nutritional quality, use of recombinant
DNA techniques, 4

O

Organ(s) for transplantation, genetic
modifications of farm animals, 225
Organic foods produced by biotechnology,
role of Agricultural Biotechnology
Research Advisory Committee, 56
Organism
definition, 25
genetically modified, safety assessment
of flavor ingredients produced, 59–67

Organization for Economic Cooperation and Development Group of National Experts on Safety in Biotechnology, genetically modified food safety assessment, 29
Oryzacystatins
  cloning, 124
  function, 124
  identification, 124
  protective effects, 124–125
Ovalbumin, role as allergen, 106
Ovomucoid, role as allergen, 106

P

Peanut, allergens, 106–107
Peas, allergens, 107
Pelargonium species, use in food quality improvement, 4
Plant
  genetic(s), safety concerns, 113–114
  genetic modification, 101
  recombinant DNA techniques, 2–4
  role in severity of viral diseases, 115
Plant breeding
  food safety testing of modified crops, 134
  role of genetic engineering techniques, 134
Plant tissue culture, production using modern biotechnology, 60–61
Plant virus genes, disease-resistant crop production, 113–122
Polyploidization, development, 114
Potato(es), viruses, 117,119
Potato(es) resistant to Colorado potato beetle, safety assessment, 148–155
Potato tuber, composition and wholesomeness, 153–155
Potyviruses
  description, 119
  role in crop loss, 119
Profilin, role as allergen, 108
Pronuclei microinjection, method of transferring genes, 211–212
Protein C, human, genetic modifications of farm animals, 223

Protein engineering
  Thermoanaerobacter cyclodextrin glycosyltransferase, 204–207
  use of recombinant DNA techniques, 5–6
Public acceptance of scientific research, role of Agricultural Biotechnology Research Advisory Committee, 55–56
Public advisory committees, role in providing advice to government on agricultural science policy, 52–57

R

Recombinant(s), safety assessment method, 44–45
Recombinant bovine somatotropin, description, 6
Recombinant chymosin, FDA regulation, 61
Recombinant DNA, ethics of use, 114
Recombinant DNA techniques
  animals, 6
  applications, 13
  description, 12–13
  enzymes, 5–6
  FDA policy, 13–21
  food ingredients, 6
  microorganisms, 4–5
  plants, 2–4
  regulation, 7–8
  safety assessment
    antibiotic resistance marker genes used in foods produced, 46–48
    food crops produced, 39–46
    importance, 6–7
    safety concerns, 13
  use
    food quality, 3–4
    herbicide tolerance, 2–3
    insect resistance, 3
    nutritional quality, 4
    production of foods and beverages, 1–2
    virus resistance, 3
Regulation, recombinant DNA techniques, 7–8
Regulatory framework for biotechnology
  horizontal legislation, 24–26
  vertical legislation, 26

Resistance marker genes, safety
evaluation, 47
Rice
allergens, 107
transgenic food, 110

S

Saccharomyces cerevisiae, recombinant
DNA techniques for food production, 5
Safety
cereal cystatins, 132
genetic modifications associated with
sucrose accumulation, 95–97
Safety assessment
acetolactate decarboxylase producing
brewer's yeast, 173–178
antibiotic resistance marker genes used
in foods produced by recombinant DNA
techniques
evaluation concerns, 47
items requiring evaluation, 48
need, 46–47
principle, 47
safety problems, 47–48
Bacillus thuringiensis insecticidal
crystal protein CRYIA(b) expressed in
transgenic tomatoes
chemical analysis of nutrients and
glycoalkaloids, 140
development of food safety test
strategy, 135
digestibility of CRYIA(b) protein upon
gastrointestinal tract passage in
rats, 138
feeding trial with transgenic
tomatoes, 141,142*f*
food safety of transgenic tomatoes,
139,142*f*
future work, 143,145
hemolytic effects of CRYIA(b)
protein, 139
in vitro binding of CRYIA(b) protein to
gastrointestinal tract tissues of
mammals, 137
in vitro degradation of CRYIA(b) and
NPRII proteins, 138

Safety assessment—*Continued*
Bacillus thuringiensis insecticidal
crystal protein CRYIA(b) expressed in
transgenic tomatoes—*Continued*
in vivo binding of CRYIA(b) protein to
gastrointestinal tract tissues of rat, 137
insecticidal activity of CRYIA(b)
protein in mammals, 136–137
production and characterization of
CRYIA(b) and NPTII proteins, 136
project description, 135
short-term toxicity of CRYIA(b) protein
in mice, 138–139
in rabbits, 139
testing protocol, 144*f*,145
toxicity profile of CRYIA(b) protein, 138
transformation and characterization of
tomato lines, 135–136
beer brewing improvement using
genetically modified yeast, 170
enzymes from genetically modified
microorganisms
cloned organism, 203
donor organism, 203
evaluation principles, 199,202–203
host organism, 203
manufacture and quality control, 199,202*t*
studies, 203–204
flavor ingredients produced by genetically
modified organisms
FDA regulation of recombinant
chymosin, 61
Flavor and Extract Manufacturers
Association GRAS program for
flavor ingredients, 61–67
flavor production using biotechnology,
60–61
Flavr Savr tomato
comparison to other tomatoes
allergenicity potential, 80
compromising antibiotic therapy
potential, 80–81
field trials, 79–80
glycoalkaloids, 79
horizontal gene transfer potential, 81
nutrients, 78–79
toxicity potential, 80

Safety assessment—*Continued*
Flavr Savr tomato—*Continued*
consultations and submissions by
Calgene to FDA and USDA, 72–74
FDA conclusions
aminoglycoside 3'-phosphotransferase
II, 82,84–85
kanamycin resistance marker gene, 85
tomato, 82,83*f*
FDA Food Advisory Committee
conclusions, 82
functions of DNA inserted, 77–78
guideline establishment, 75
identity of tomatoes, 75–76
food crops produced by recombinant DNA
techniques
evaluation method
basis, 41–42
food or additive identical with host, 42
introduced gene, 42–44
principles, 40
recombinants, 44–45
vector, 42
functions of introduced gene, 46
necessary information, 45–46
need, 39–40
products covered, 40
genetically engineered crops
Flavr Savr tomato, 72–85
guideline establishment, 75
genetically modified foods, 29–31
genetically modified lactic acid bacteria
characteristics of food-grade systems,
185–186
factors affecting conjugal transfer,
188–193
food-grade marker gene, 187
food-grade vectors, 187–189*f*
foreign marker genes, 187–193
host of foreign gene, 186–187
lactic acid bacteria
relation to human life, 182–183
role in food fermentation, 183–185
potatoes resistant to Colorado potato beetle
allergenic potential 155
Bacillus thuringiensis subsp.
tenebrionis protein, 152–153

Safety assessment—*Continued*
potatoes resistant to Colorado potato
beetle—*Continued*
composition and wholesomeness of
potato tuber, 153–155
donor organisms and transformation
process, 150*f*,151
genetic modification of potato
plants, 149
NPTII protein, 153
procedure, 149
protein expression levels, 151–152
recombinant DNA techniques, 6–7
Scandinavian countries, scientific
advisory committees for genetically
modified food safety assessment, 30
Scientific advisory committees,
genetically modified food safety
assessment, 30–31
Scientific Issues Related to Potential
Allergenicity in Transgenic Food Crops
conference, goal, 17
Scientific research, role of Agricultural
Biotechnology Research Advisory
Committee in public acceptance, 56
Screening marker genes, function, 47
Second-generation antiviral
technologies, effectiveness
and safety, 122
Selection cassette, presence and
expression in virus-resistant
transgenic plants, 121
Selective marker genes, function, 47
Soy products, use as meat substitute,
160
Soybean, modified, transgenic food,
109–110
Standard of care, description, 15
Stevioside, glycosylation, 197,198*f*
Substantial equivalence, description,
7,149
Sucrose accumulation in tomato
genetic accumulation, 95,96*f*
introgression, 90–94
molecular cloning of responsible gene,
93–95
safety of genetic modifications, 95–97

T

Thermoanaerobacter cyclodextrin glycosyltransferase
cyclodextrin production, 197,204,206*f*
development, 197,199–201*f*
properties, 204
protein engineering
β-cyclodextrin crystallization, 205,206*f*
organic solvent complexation, 205,207*f*
Tissue plasminogen activator, genetic modifications of farm animals, 223
Tomato
effect of genetic modification on carbohydrate composition, 89
transgenic food, 109
Tomato lines, transformation and characterization, 135–136
Transformation, description, 12–13
Transgene, definition, 210
Transgenic animals
definition, 210
use for food and fiber, 211
Transgenic *Bacillus thuringiensis* tomatoes
feeding trial, 141,142*f*
food safety, 139,142*f*
Transgenic foods
allergenicity, 100–110
Bacillus thuringiensis cotton seed, 109
Flavr Savr tomato, 108–109
modified soybean, 109–110
rice, 110
transgenic tomatoes, 109
Transgenic rice cultivar, construction, 131
Transgenic tomatoes, safety assessment of *Bacillus thuringiensis* insecticidal crystal protein CRYIA(b), 134–145

U

United Kingdom, scientific advisory committees for genetically modified food safety assessment, 30
United States Department of Agriculture (USDA), roles in agricultural biotechnology, 53

V

Vector
role in severity of viral diseases, 115
safety assessment method, 42
Vertical legislation, concept, 26
Viral diseases, components determining severity, 115,116*f*
Virus
content in tomatoes, 114–115
role in severity of viral diseases, 115,117,118*f*
Virus resistance
experimental protocol for testing, 119–121
use of recombinant DNA techniques, 3
Virus-resistant transgenic plants, presence and expression of selection cassette, 121

W

Wheat, allergens, 107
Whey acidic protein, genetic modifications of farm animals, 222–223

Y

Yeast, genetically modified, beer brewing improvement, 160–170

Bestsellers from ACS Books

The ACS Style Guide: A Manual for Authors and Editors
Edited by Janet S. Dodd
264 pp; clothbound ISBN 0–8412–0917–0; paperback ISBN 0–8412–0943–X

Understanding Chemical Patents: A Guide for the Inventor
By John T. Maynard and Howard M. Peters
184 pp; clothbound ISBN 0–8412–1997–4; paperback ISBN 0–8412–1998–2

Chemical Activities (student and teacher editions)
By Christie L. Borgford and Lee R. Summerlin
330 pp; spiralbound ISBN 0–8412–1417–4; teacher ed. ISBN 0–8412–1416–6

Chemical Demonstrations: A Sourcebook for Teachers,
Volumes 1 and 2, Second Edition
Volume 1 by Lee R. Summerlin and James L. Ealy, Jr.;
Vol. 1, 198 pp; spiralbound ISBN 0–8412–1481–6;
Volume 2 by Lee R. Summerlin, Christie L. Borgford, and Julie B. Ealy
Vol. 2, 234 pp; spiralbound ISBN 0–8412–1535–9

Chemistry and Crime: From Sherlock Holmes to Today's Courtroom
Edited by Samuel M. Gerber
135 pp; clothbound ISBN 0–8412–0784–4; paperback ISBN 0–8412–0785–2

Writing the Laboratory Notebook
By Howard M. Kanare
145 pp; clothbound ISBN 0–8412–0906–5; paperback ISBN 0–8412–0933–2

Developing a Chemical Hygiene Plan
By Jay A. Young, Warren K. Kingsley, and George H. Wahl, Jr.
paperback ISBN 0–8412–1876–5

Introduction to Microwave Sample Preparation: Theory and Practice
Edited by H. M. Kingston and Lois B. Jassie
263 pp; clothbound ISBN 0–8412–1450–6

Principles of Environmental Sampling
Edited by Lawrence H. Keith
ACS Professional Reference Book; 458 pp;
clothbound ISBN 0–8412–1173–6; paperback ISBN 0–8412–1437–9

Biotechnology and Materials Science: Chemistry for the Future
Edited by Mary L. Good (Jacqueline K. Barton, Associate Editor)
135 pp; clothbound ISBN 0–8412–1472–7; paperback ISBN 0–8412–1473–5

For further information and a free catalog of ACS books, contact:
American Chemical Society
Product Services Office
1155 16th Street, NW, Washington, DC 20036
Telephone 800–227–5558